Applications of
Abstract Algebra
with MAPLE

Applications of Abstract Algebra with MAPLE

Richard E. Klima
Neil Sigmon
Ernest Stitzinger

CRC Press
Boca Raton   London   New York   Washington, D.C.

Library of Congress Cataloging-in-Publication Data

Klima, Richard E.
    Applications of abstract algebra with Maple / Richard E. Klima,
Neil P. Sigmon, Ernest Stitzinger.
       p. cm.— (Discrete mathematics and its applications)
    Includes bibliographical references and index.
    ISBN 0-8493-8170-3 (alk. paper)
    1. Algebra, Abstract—Data processing. 2. Maple (Computer file)
I. Stitzinger, Ernest. II. Sigmon, Neil P. III. Title.
IV. Series.
QA162.K65 1999
512′.02′02855369—dc21
                                   99-37392
                                     CIP

Preface

In 1990 we introduced a one-semester applications of algebra course at North Carolina State University for students who had successfully completed semesters of linear and abstract algebra. We intended for the course to give students more exposure to basic algebraic concepts, and to show students some practical uses of these concepts. The course was received enthusiastically by both students and faculty and has become one of the most popular mathematics electives at NC State.

When we were originally deciding on material for the course, we knew that we wanted to include several topics from coding theory, cryptography, and counting (what we call *Polya* theory). With this in mind, at the suggestion of Michael Singer, we used George Mackiw's book, *Applications of Abstract Algebra*, for the first few years, and supplemented as we saw fit. After several years, Mackiw's book went out of print temporarily. Rather than search for a new book for the course, we decided to write our own notes and teach the course from a coursepack. About the same time, NC State incorporated the mathematics software package Maple V^{TM}[1] into its calculus sequence, and we decided to incorporate it into our course as well. The use of Maple played a central role in the recent development of the course because it provides a way for students to see realistic examples of the topics discussed without having to struggle with extensive computations. With additional notes regarding the use of Maple in the course, our coursepack evolved into this book. In addition to the topics discussed in this book, we have included a number of other topics in the course. However, the present material has become the constant core for the course.

Our philosophy concerning the use of technology in the course is that it be a useful tool and not present new problems or frustrations. Consequently, we have included very detailed instructions regarding the use of

---

[1] Maple V is a registered trademark of Waterloo Maple, Inc., 57 Erb St. W, Waterloo, Canada N2L6C2, www.maplesoft.com.

Maple in this book. It is our hope that the Maple discussions are thorough enough to allow it to be used without much alternative aid. As alternative aids, we have included a basic Maple tutorial in Appendix A, and an introduction to some of Maple's linear algebra commands in Appendix B. Although we do not require students to produce the Maple code used in the course, we do require that they obtain a level of proficiency such that they can make basic changes to provided worksheets to complete numerous Maple exercises. So that this book can be used for applications of algebra courses in which Maple is not incorporated, we have separated all Maple material into sections that are clearly labeled, and separated all Maple and non-Maple exercises.

When teaching the course, we discuss the material in Chapter 1 as needed rather than review it all at once. More specifically, we discuss the material in Chapter 1 through examples the first time it is needed in the applications that follow. Some of the material in Chapter 1 is review material that does not apply specifically to the applications that follow. However, for students with weak backgrounds, Chapter 1 provides a comprehensive review of all necessary prerequisite mathematics.

Chapter 2 is a short chapter on block designs. In Chapters 3, 4, and 5 we discuss some topics from coding theory. In Chapter 3 we introduce error-correcting codes, and present Hadamard, Reed-Muller, and Hamming codes. In Chapters 4 and 5, we present BCH codes and Reed-Solomon codes. Each of these chapters are dependent in part on the preceding chapters. The dependency of Chapter 3 on Chapter 2 can be avoided by omitting Sections 3.2, 3.3, and 3.4 on Hadamard and Reed-Muller codes. In Chapters 6, 7, and 8 we discuss some topics from cryptography. In Chapter 6 we introduce algebraic cryptography, and present several variations of the Hill cryptosystem. In Chapter 7 we present the RSA cryptosystem and discuss some related topics, including the Diffie-Hellman key exchange. In Chapter 8 we present the ElGamal cryptosystem, and describe how elliptic curves can be incorporated into the system naturally. There is a slight dependency of Chapters 7 and 8 on Chapter 6, and of Chapter 8 on Chapter 7. Chapter 9 is a stand-alone chapter in which we discuss the Polya counting techniques, including Burnside's Theorem and the Polya Enumeration Theorem.

We wish to thank all those who have been involved in the development of this course and book. Pete Hardy taught from the coursepack and improved it with his suggestions. Also, Michael Singer suggested various topics and wrote notes on some of them. Many students have written on this material for various projects. Of these, the recent master's project by Karen Klein on elliptic curves was especially interesting. Finally, we wish to thank our mentor, Jack Levine, for his interest in our projects, his guidance as we learned about applications of algebra, and his many contributions to the subject, especially in cryptography.

Contents

Chapter 1

Preliminary Mathematics

There are two purposes to this chapter. We very quickly and concisely review some of the basic algebraic concepts that are probably familiar to many readers, and also introduce some topics for specific use in later chapters. We will generally not pursue topics any further than is necessary to obtain the material needed for the applications that follow. Topics discussed in this chapter include permutation groups, the ring of integers, polynomial rings, finite fields, and examples that incorporate these topics using the philosophies of concepts covered in later chapters.

1.1 Permutation Groups

Suppose a set G is *closed* under an operation $*$. That is, suppose $a * b \in G$ for all $a, b \in G$. Then $*$ is called a *binary operation* on G. We will use the notation $(G, *)$ to represent the set G with this operation. Suppose $(G, *)$ also satisfies the following three properties.

1. $(a * b) * c = a * (b * c)$ for all $a, b, c \in G$.

2. There exists an *identity* element $e \in G$ for which $e * a = a * e = a$ for all $a \in G$.

3. For each $a \in G$, there exists an *inverse* element $b \in G$ for which $a * b = b * a = e$. The inverse of a is usually denoted a^{-1} or $-a$.

Then $(G, *)$ is called a *group*. For example, it can easily be verified that for the set Z of integers, $(Z, +)$ is a group with identity element 0.

Let S be a set, and let $A(S)$ be the set of bijections on S. Then an element $\alpha \in A(S)$ can be uniquely expressed by its action $(s)\alpha$ on the elements $s \in S$.

Example 1.1 If $S = \{1, 2, 3\}$, then $A(S)$ contains six elements. One of the α in $A(S)$ can be expressed as $(1)\alpha = 2$, $(2)\alpha = 3$, and $(3)\alpha = 1$. ■

Let $\circ$ represent the composition operation on $A(S)$. Specifically, if $\alpha, \beta \in A(S)$, then define $\alpha \circ \beta$ by the action $(s)(\alpha \circ \beta) = ((s)\alpha)\beta$ for $s \in S$. Since the composition of two bijections on S is also a bijection on S, then $\alpha \circ \beta \in A(S)$. Hence, $\circ$ is a binary operation on $A(S)$. It can easily be verified that $(A(S), \circ)$ is a group (see Written Exercise 1).

A group $(G, *)$ is said to be *abelian* or *commutative* if $a * b = b * a$ for all $a, b \in G$. For example, since $m + n = n + m$ for all $m, n \in Z$, then $(Z, +)$ is abelian. However, for a set S with more than two elements, $\alpha \circ \beta \neq \beta \circ \alpha$ for some $\alpha, \beta \in A(S)$. Therefore, if a set S contains more than two elements, then $(A(S), \circ)$ is not abelian.

We will represent the number of elements in a set S by $|S|$. Suppose S is a set with $|S| = n$. Then $(A(S), \circ)$ is denoted by S_n and called the *symmetric group on n letters*. It can easily be shown that $|S_n| = n!$ (see Written Exercise 2). Suppose $\alpha \in S_n$. Then α can be viewed as a bijection on the set $\{1, 2, \ldots, n\}$. This bijection can be represented by listing the elements in the set $\{1, 2, \ldots, n\}$ in a row with their images under α listed immediately below.

$$\alpha : \begin{pmatrix} 1 & 2 & \cdots & n \\ (1)\alpha & (2)\alpha & \cdots & (n)\alpha \end{pmatrix}$$

Example 1.2 Let $S = \{1, 2, 3\}$, and let $\alpha \in S_3$ be given by $(1)\alpha = 2$, $(2)\alpha = 3$, and $(3)\alpha = 1$. Then α can be represented as follows.

$$\alpha : \begin{pmatrix} 1 & 2 & 3 \\ 2 & 3 & 1 \end{pmatrix}$$

■

An element $\alpha \in S_n$ is called a *permutation*. Note that for permutations $\alpha, \beta \in S_n$, we can represent $\alpha \circ \beta$ as follows.

$$\begin{pmatrix} 1 & \cdots & n \\ (1)\alpha & \cdots & (n)\alpha \end{pmatrix} \begin{pmatrix} 1 & \cdots & n \\ (1)\beta & \cdots & (n)\beta \end{pmatrix} = \begin{pmatrix} 1 & \cdots & n \\ (1\alpha)\beta & \cdots & (n\alpha)\beta \end{pmatrix}$$

For example, let $\alpha \in S_4$ be given by $(1)\alpha = 2$, $(2)\alpha = 4$, $(3)\alpha = 3$, and $(4)\alpha = 1$, and let $\beta \in S_4$ be given by $(1)\beta = 4$, $(2)\beta = 3$, $(3)\beta = 2$, and

$(4)\beta = 1$. Then we can express $\alpha \circ \beta$ as follows.

$$\begin{pmatrix} 1 & 2 & 3 & 4 \\ 2 & 4 & 3 & 1 \end{pmatrix} \begin{pmatrix} 1 & 2 & 3 & 4 \\ 4 & 3 & 2 & 1 \end{pmatrix} = \begin{pmatrix} 1 & 2 & 3 & 4 \\ 3 & 1 & 2 & 4 \end{pmatrix}$$

We now discuss another way to express elements in S_n. Let $i_1, i_2, \ldots, i_s$ be distinct elements in the set $S = \{1, 2, \ldots, n\}$. Then $(i_1\ i_2\ i_3\ \cdots\ i_{s-1}\ i_s)$ is called a *cycle of length s* or an *s-cycle*, and represents the permutation $\alpha \in S_n$ that maps $i_1 \mapsto i_2, i_2 \mapsto i_3, \ldots, i_{s-1} \mapsto i_s, i_s \mapsto i_1$, and every other element in S to itself. For example, the permutation

$$\alpha : \begin{pmatrix} 1 & 2 & 3 & 4 & 5 & 6 \\ 3 & 4 & 5 & 1 & 6 & 2 \end{pmatrix}$$

in S_6 can be expressed as the 6-cycle (135624). Note that this expression of α as a cycle is not unique, for α can also be expressed as (356241) and (562413), among others.

Next, consider the permutation

$$\beta : \begin{pmatrix} 1 & 2 & 3 & 4 & 5 & 6 \\ 3 & 4 & 5 & 6 & 1 & 2 \end{pmatrix}$$

in S_6. To express β using cycle notation, we must use more than one cycle. For example, we can express β as the following "product" of two 3-cycles: (135)(246). Since these cycles contain no elements in common they are said to be *disjoint*. And because they are disjoint, the order in which they are listed does not matter. The permutation β can also be expressed as (246)(135).

Every permutation in S_n can be expressed as either a single cycle or a product of disjoint cycles. When a permutation is expressed as a product of disjoint cycles, cycles of length one are not usually included. For example, consider the permutation

$$\gamma : \begin{pmatrix} 1 & 2 & 3 & 4 & 5 & 6 \\ 3 & 4 & 5 & 2 & 1 & 6 \end{pmatrix}$$

in S_6. Even though the fact that γ maps 6 to itself would be expressed as the 1-cycle (6), this cycle would not usually be included in the expression of γ as a product of disjoint cycles. That is, γ would usually be expressed as (135)(24) or (24)(135).

In an expression of a permutation as a product of cycles, the cycles need not be disjoint. For example, the permutation $\alpha = (135624)$ defined above can also be expressed as the product (13)(15)(16)(12)(14) of 2-cycles.

Because these 2-cycles are not disjoint, the order in which they are listed matters.

A 2-cycle is also called a *transposition*. Any permutation can be expressed as a product of transpositions in the way illustrated above for α. Specifically, the cycle $(i_1 \, i_2 \, i_3 \, \cdots \, i_{s-1} \, i_s)$ can be expressed as the product $(i_1 \, i_2)(i_1 \, i_3) \, \cdots \, (i_1 \, i_{s-1})(i_1 \, i_s)$ of transpositions. If a permutation can be expressed as a product of more than one disjoint cycle, then each cycle can be considered separately when expressing the permutation as a product of transpositions. For example, the permutation $\beta = (135)(246)$ defined above can be expressed as $(13)(15)(24)(26)$, and the permutation $\gamma = (135)(24)$ defined above can be expressed as $(13)(15)(24)$.

There are many ways to express a permutation as a product of transpositions, and the number of transpositions in these expressions may vary. However, the number of transpositions in the expression of a permutation as a product of transpositions is either always even or always odd. A permutation is said to be *even* if it can be expressed as a product of an even number of transpositions, and *odd* if it can be expressed as a product of an odd number of transpositions. Thus, the product of two even permutations is even, and the product of two odd permutations is also even.

The inverse of the cycle $(i_1 \, i_2 \, i_3 \, \cdots \, i_{s-1} \, i_s)$ is $(i_s \, i_{s-1} \, \cdots \, i_3 \, i_2 \, i_1)$. Suppose $\alpha = \alpha_1 \alpha_2 \cdots \alpha_k \in S_n$, where each α_i is a transposition. Then $\alpha^{-1} = \alpha_k^{-1} \cdots \alpha_2^{-1} \alpha_1^{-1} = \alpha_k \cdots \alpha_2 \alpha_1$ since $\alpha_i^{-1} = \alpha_i$ for each transposition α_i. Hence, the inverse of an even permutation is even. And because the identity permutation is even, the subset of even permutations in S_n forms a group. This group is denoted by A_n and called the *alternating group on n letters*. Since A_n is a subset of S_n and forms a group, we call A_n a *subgroup* of S_n.

Definition 1.1 *Let $(G, *)$ be a group, and suppose H is a nonempty subset of G. If $(H, *)$ is a group, then H is called a subgroup of G.*

Consider a regular polygon P, such as, for example, an equilateral triangle or a square. Any movement of P that preserves the general shape of P is called a *rigid motion*. There are two types of rigid motions – rotations and reflections. For a regular polygon P with n sides, there are $2n$ distinct rigid motions. These include the n rotations of P through $360j/n$ degrees for $j = 1, \ldots, n$. The remaining n rigid motions are reflections. If n is even, these are the reflections of P across the lines that connect opposite vertices or bisect opposite sides of P. If n is odd, these are the reflections of P across the lines that are perpendicular bisectors of the sides of P. Since the rigid motions of P preserve the general shape of P, they can be viewed as permutations of the vertices or sides of P. The set of rigid motions of a regular polygon P forms a group called the *symmetries* of P.

Example 1.3 Consider the group of symmetries of a square. To express these symmetries as permutations of the vertices of a square, consider the following general figure.

The 8 symmetries of a square can be expressed as permutations of the vertices of this general figure as follows (rotations are counterclockwise).

Rigid Motion	Permutation
90° rotation	(1234)
180° rotation	(13)(24)
270° rotation	(1432)
360° rotation	identity
reflection across horizontal	(12)(34)
reflection across vertical	(14)(23)
reflection across 1–3 diagonal	(24)
reflection across 2–4 diagonal	(13)

Note that expressing these rigid motions as permutations on the vertices of the preceding general figure yields a subgroup of S_4. ∎

When the symmetries of an n-sided regular polygon are expressed as permutations on the set $\{1, 2, \ldots, n\}$, the resulting subgroup of S_n is denoted by D_n and called the *dihedral group on n letters*. The subgroup of S_4 in Example 1.3 is the dihedral group D_4.

A group $(G, \cdot)$, or just G for short, is called *cyclic* if there is an element $a \in G$ for which $G = \{a^i \mid i \in Z\}$. In this case, a is called a *cyclic generator* for G. More generally, suppose a is an element in a group G, and let $H = \{a^i \mid i \in Z\}$. Then H is a subgroup of G called the *cyclic group generated by* a. Let $a^i = a^j$ for some $0 < i < j$. Then $a^{j-i} = a^j a^{-i} = e$, where e is the identity element in G. Thus, there is a smallest positive integer m for which $a^m = e$. Now, suppose $a^t = e$. Since $t = mq + r$ for some $0 \le r < m$, and $a^t = a^{mq+r} = (a^m)^q a^r = a^r$, it follows that $r = 0$. Hence, m divides t. Since $a^i = a^j$ for $i < j$ forces $a^{j-i} = e$, a contradiction if $0 < j - i < m$, the set $\{a^i \mid 0 \le i < m\}$ consists of m

distinct elements. Furthermore, for any integer k we can write $k = mq + r$ for some $0 \leq r < m$ with $a^k = a^r$. Therefore, $H = \{a^i \mid 0 \leq i < m\}$, and H contains m elements. We summarize this discussion as the following theorem.

Theorem 1.2 *Suppose a is an element in a group G. If m is the smallest positive integer for which $a^m = e$, where e is the identity element in G, then the cyclic group generated by a contains m elements.*

The value of m in Theorem 1.2 is called the *order* of a. Also, a set S with $|S| = n$ is said to have *order* n. Hence, the order of an element a in a group G is the order of the cyclic subgroup of G generated by a. We will show in Theorem 1.4 that for an element of order m in a group G of order n, m must divide n. Therefore, in a group G of order n, $a^n = e$ for all $a \in G$ where e is the identity element in G. We summarize this as the following corollary.

Corollary 1.3 *Suppose a is an element in a group G of order n. Then $a^n = e$ where e is the identity element in G.*

Example 1.4 Consider the dihedral group D_n of order $2n$. Recall that the elements in D_n can be viewed as the symmetries of an n-sided regular polygon P. Each of the n reflections of P has order 2. Also, the rotations of P through $360/n$ and $360(n-1)/n$ degrees have order n (as do, possibly, some other rotations). Note that these orders divide $|D_n|$.   ∎

1.2   Cosets and Quotient Groups

Let H be a subgroup of a group G. For an element $g \in G$, we define $gH = \{gh \mid h \in H\}$, called a *left coset* of H in G. Since $gh_1 = gh_2$ implies $h_1 = h_2$ for all $h_1, h_2 \in H$, then there is a one-to-one correspondence between the elements in gH and H. Thus, if H is finite, $|gH| = |H|$. Suppose $g_1, g_2 \in G$. If $x \in g_1H \cap g_2H$ for some $x \in G$, then $x = g_1h_1 = g_2h_2$ for some $h_1, h_2 \in H$. Hence, $g_1 = g_2h_2h_1^{-1} \in g_2H$. Then for any $y \in g_1H$, it follows that $y = g_1h_3 = g_2h_2h_1^{-1}h_3 \in g_2H$ for some $h_3 \in H$. Therefore, $g_1H \subseteq g_2H$. Similarly, $g_2H \subseteq g_1H$, so $g_1H = g_2H$. The preceding arguments imply that if $g_1, g_2 \in G$, then either $g_1H = g_2H$, or g_1H and g_2H are disjoint. Hence, G is the union of pairwise disjoint left cosets of H in G.

Example 1.5 Consider the subgroup A_n of S_n. If α is an odd permutation in S_n, then αA_n and A_n are disjoint. If β is any other odd permutation in S_n, then $\beta^{-1}\alpha$ will be even. Therefore, $\beta^{-1}\alpha \in A_n$, and $\alpha A_n = \beta A_n$. Hence, there are two left cosets of A_n in S_n, one consisting of the even permutations in S_n, and the other consisting of the odd permutations. ∎

For a finite group G with subgroup H, the following theorem is a fundamental algebraic result regarding the number of left cosets of H in G. This theorem is called *Lagrange's Theorem*.

Theorem 1.4 *Let G be a group of order n with subgroup H of order k, and suppose there are t distinct left cosets of H in G. Then $n = kt$.*

Proof. Each of the t distinct left cosets of H in G contains k elements. Since G is the union of these left cosets, then $n = kt$. ∎

As a consequence of Lagrange's Theorem, the order of a subgroup H in a finite group G must divide the order of G. For example, the dihedral group D_4 of permutations in Example 1.3 has order 8, which divides $|S_4| = 24$.

We began this section by defining the left cosets gH of a subgroup H in a group G. Results analogous to those discussed so far in this section also hold for the sets $Hg = \{hg \mid h \in H\}$, called *right cosets* of H in G.

Next, we discuss how cosets can be used to construct new groups from known ones. Suppose H is a subgroup of a group G. Then for $x \in G$, let $x^{-1}Hx = \{x^{-1}hx \mid h \in H\}$. If $x^{-1}Hx \subseteq H$ for all $x \in G$, then H is called a *normal* subgroup of G. As we will show, if H is a normal subgroup of a group G, then the set of left cosets of H in G forms a group with the operation $(xH)(yH) = (xy)H$. To see this, note first that since H is normal in G, then $x^{-1}Hx \subseteq H$ for all $x \in G$. Specifically, this will be true if we replace x with x^{-1}. That is, $(x^{-1})^{-1}Hx^{-1} = xHx^{-1} \subseteq H$. Thus, for any $h \in H$, it follows that $h = x^{-1}(xhx^{-1})x = x^{-1}h_1x \in x^{-1}Hx$ for some $h_1 \in H$. Hence, $H \subseteq x^{-1}Hx$, and since H is normal in G, then $x^{-1}Hx = H$. Therefore, a subgroup H in a group G satisfies $x^{-1}Hx = H$ if and only if H is normal in G.

To see that the operation defined above for the left cosets of H in G is well-defined, let $xH = x_1H$ and $yH = y_1H$ for some $x, x_1, y, y_1 \in G$. Since $xH = x_1H$ and $yH = y_1H$, then $x = x_1h_1$ and $y = y_1h_2$ for some $h_1, h_2 \in H$. And since H is normal in G, then $y_1^{-1}h_1y_1 = h_3$ for some $h_3 \in H$, or, equivalently, $h_1y_1 = y_1h_3$ for some $h_3 \in H$. This yields $xy = x_1h_1y_1h_2 = x_1y_1h_3h_2 \in x_1y_1H$. Thus, $xy \in x_1y_1H$, and $xyH = x_1y_1H$.

Therefore, the operation defined above for the left cosets of H in G is well-defined.

We can now easily show that if H is a normal subgroup of a group G, then the set of left cosets of H in G forms a group with the operation $(xH)(yH) = (xy)H$. This group, denoted G/H, is called a *quotient group*.

Theorem 1.5 *Suppose H is a normal subgroup of a group G. Then the set $G/H = \{xH \mid x \in G\}$ of left cosets of H in G forms a group with the operation $(xH)(yH) = (xy)H$.*

Proof. If e is the identity element in G, then $eH = H$ is the identity in G/H since $(eH)(xH) = (ex)H = xH$ and $(xH)(eH) = (xe)H = xH$ for all $x \in G$. Also, the inverse of the element xH in G/H is $x^{-1}H$ since $(x^{-1}H)(xH) = (x^{-1}x)H = eH = H$. The associative law in G/H can easily be verified. ∎

Note that if G is abelian, then any subgroup H of G is normal and G/H is abelian.

Example 1.6 Let $G = (Z, +)$. Choose an integer $n \in Z$, and let H be the cyclic subgroup of G generated by n. Since the operation on this group is addition, then $H = \{kn \mid k \in Z\}$ and additive notation $x + H$ is used for the cosets of H in G. That is, the cosets of H in G are the sets $x + H = \{x + h \mid h \in H\} = \{x + kn \mid k \in Z\}$ for all $x \in Z$. The distinct left cosets of H in G are the sets $H, 1 + H, 2 + H, \ldots, (n-1) + H$. Hence, G/H consists of these sets with the operation $(x + H) + (y + H) = (x + y) + H$. Note that if we would perform this operation without including H in the notation, we would simply be doing integer addition modulo n. Note also that G/H is cyclic with generator $1 + H$. ∎

Suppose H is a normal subgroup of a group G, and define the mapping $\varphi : G \to G/H$ by $\varphi(x) = xH$. For this mapping φ, it can easily be seen that $\varphi(xy) = \varphi(x)\varphi(y)$ for all $x, y \in G$. Since φ satisfies this property, we call φ a *homomorphism*.

Definition 1.6 *Let G and H be groups. A mapping $\varphi : G \to H$ that satisfies $\varphi(xy) = \varphi(x)\varphi(y)$ for all $x, y \in G$ is called a homomorphism.*

Example 1.7 Let H be the group $H = \{odd, even\}$ with identity element *even*. Define $\varphi : S_n \to H$ by $\varphi(x) = even$ if x is an even permutation, and $\varphi(x) = odd$ if x is an odd permutation. Then φ is a homomorphism. ∎

Example 1.8 Let G be the multiplicative group of nonsingular $n \times n$ matrices over the reals (*i.e.*, with entries in the reals). Then the determinant function is a homomorphism from G onto the multiplicative group of nonzero reals. ∎

Let φ be a homomorphism from G into H. We define the *kernel* of φ to be the set Ker $\varphi = \{g \in G \mid \varphi(g) = e\}$, where e is the identity element in H. It can easily be verified that Ker φ is a normal subgroup of G (see Written Exercise 14). Also, if H is a normal subgroup of G, and if we define the mapping $\varphi : G \to G/H$ by $\varphi(x) = xH$, then Ker $\varphi = H$. Hence, every normal subgroup of a group G is the kernel of a homomorphism with domain G, and the kernel of every homomorphism with domain G is a normal subgroup of G.

1.3 Rings and Euclidean Domains

Let R be a set with two binary operations, an addition "+" and multiplication "$*$". Suppose R also satisfies the following three properties.

1. $(R, +)$ is an abelian group with identity element we will denote by 0.

2. $(a * b) * c = a * (b * c)$ for all $a, b, c \in R$.

3. $a * (b + c) = (a * b) + (a * c)$ and $(a + b) * c = (a * c) + (b * c)$ for all $a, b, c \in R$.

Then R is called a *ring*. If also $a * b = b * a$ for all $a, b \in R$, then R is said to be *commutative*. And if there exists a multiplicative identity element $1 \in R$ for which $1 * a = a * 1 = a$ for all $a \in R$, then R is said to be a *ring with identity*. As is customary, we will suppress the $*$ from the notation when performing the multiplication operation in a ring.

All of the rings we will use in this book will be commutative with identity. A commutative ring R with identity is called an *integral domain* if $ab = 0$ with $a, b \in R$ implies $a = 0$ or $b = 0$. A commutative ring R with identity is called a *field* if every nonzero element in R has a multiplicative inverse in R. All fields are integral domains.

Two rings we will use extensively are the ring $F[x]$ of polynomials in x with coefficients in a field F, and the ring Z of integers with the usual operations of addition and multiplication. Both $F[x]$ and Z are integral domains, but not fields.

Suppose B is a nonempty subset of a commutative ring R. If $(B, +)$ is a subgroup of $(R, +)$, and if $rb \in B$ for all $r \in R$ and $b \in B$, then B is called an *ideal* of R. If also there exists an element $b \in B$ for which $B = \{rb \mid r \in R\}$, then B is called a *principal ideal*. In this case we denote $B = (b)$ and call B the *ideal generated by b*.

If $f(x) \in F[x]$, then $(f(x))$ consists of all multiples of $f(x)$ over F. That is, $(f(x))$ consists of all polynomials in $F[x]$ of which $f(x)$ is a factor. A similar result holds for integers $n \in Z$. We will show in Theorem 1.9 that all ideals in $F[x]$ and Z are principal ideals.

Ideals play a role in ring theory analogous to the role played by normal subgroups in group theory. For example, we can use an ideal of a known ring to construct a new ring. Suppose B is an ideal in a commutative ring R. Since $(B, +)$ is a subgroup of the abelian group $(R, +)$, it follows that $R/B = \{r + B \mid r \in R\}$ is an abelian group with the addition operation $(r + B) + (s + B) = (r + s) + B$. In fact, R/B is a commutative ring with the multiplication operation $(r + B)(s + B) = (rs) + B$. To see that this multiplication operation is well-defined, let $r + B = r_1 + B$ and $s + B = s_1 + B$ for some $r, r_1, s, s_1 \in R$. Since $r + B = r_1 + B$ and $s + B = s_1 + B$, then $r = r_1 + b_1$ and $s = s_1 + b_2$ for some $b_1, b_2 \in B$. But $rs = (r_1 + b_1)(s_1 + b_2) = r_1 s_1 + r_1 b_2 + b_1 s_1 + b_1 b_2 \in r_1 s_1 + B$. Thus, $rs \in r_1 s_1 + B$, and hence, $rs + B = r_1 s_1 + B$. Therefore, the multiplication operation defined above for R/B is well-defined. The ring R/B is called a *quotient ring*.

Suppose B is an ideal in a commutative ring R, and we define the mapping $\varphi : R \rightarrow R/B$ by $\varphi(x) = x + B$. For this mapping φ, it can easily be seen that $\varphi(rs) = \varphi(r)\varphi(s)$ and $\varphi(r + s) = \varphi(r) + \varphi(s)$ for all $r, s \in R$. Since φ satisfies these properties, we call φ a *ring homomorphism*.

Definition 1.7 *Let R and S be rings. A mapping $\varphi : R \rightarrow S$ that satisfies $\varphi(rs) = \varphi(r)\varphi(s)$ and $\varphi(r + s) = \varphi(r) + \varphi(s)$ for all $r, s \in R$ is called a ring homomorphism. We define the kernel of φ as $Ker \ \varphi = \{r \in R \mid \varphi(r) = 0\}$.*

Proposition 1.8 *Let R and S be commutative rings, and suppose φ is a ring homomorphism from R onto S. Then the following statements hold.*

1. *If B is an ideal in R, then the set $\varphi(B) = \{\varphi(r) \in S \mid r \in B\}$ is an ideal in S.*

2. *If B is an ideal in S, then the set $\varphi^{-1}(B) = \{r \in R \mid \varphi(r) \in B\}$ is an ideal in R.*

Proof. Exercise.   ∎

If every ideal in an integral domain D is a principal ideal, then D is called a *principal ideal domain*.

We will represent the nonzero elements in a set S by S^*. Let D be an integral domain, and let N be the set of nonnegative integers. Suppose there is a mapping $\delta : D^* \to N$ such that for $a \in D$ and $b \in D^*$, there exists $q, r \in D$ for which $a = bq + r$ with $r = 0$ or $\delta(r) < \delta(b)$. Then D is called a *Euclidean domain*. Two examples of Euclidean domains are the ring $F[x]$ of polynomials over a field F with $\delta(f(x)) = \deg f(x)$, and the ring Z of integers with $\delta(n) = |n|$.

Theorem 1.9 *Suppose D is a Euclidean domain. Then D is a principal ideal domain.*

Proof. Let B be a nonzero ideal in D, and let $b \in B$ such that $\delta(b)$ is the minimum of all $\delta(x)$ with $x \in B$. Then choose $a \in B$. Since D is a Euclidean domain, there exists $q, r \in D$ such that $a = bq + r$ with $r = 0$ or $\delta(r) < \delta(b)$. But since $r = a - bq$ and B is an ideal, then $r \in B$. By the choice of b, it follows that $r = 0$. Therefore, $a = bq$, and $a \in (b)$. Hence, $B \subseteq (b)$, but certainly $(b) \subseteq B$, so $B = (b)$. ∎

If an element a in an integral domain D has a multiplicative inverse in D, then a is called a *unit*. We will denote the set of units in an integral domain D by $U(D)$. For example, $U(Z) = \{1, -1\}$, and $U(F[x]) = \{f(x) \mid f(x)$ is a nonzero constant in $F\}$. Elements $a, b \in D$ are called *associates* if $a = ub$ for some unit $u \in D$. The only associates of an element $n \in Z$ are n and $-n$. The associates of a polynomial $f(x) \in F[x]$ are $cf(x)$ for any nonzero $c \in F$.

For elements a and b in an integral domain D, suppose there exists $x \in D$ for which $ax = b$. Then a is said to *divide* b, written $a|b$.

Proposition 1.10 *Let a, b, and c be elements in an integral domain D. Then the following statements hold.*

1. *If $a|b$ and $b|c$, then $a|c$.*

2. *$a|b$ and $b|a$ if and only if a and b are associates in D.*

3. *$a|b$ if and only if $(b) \subseteq (a)$.*

4. *$(a) = (b)$ if and only if a and b are associates in D.*

Proof. Exercise. ∎

A nonzero element a in a Euclidean domain D is said to be *irreducible* if for all $b \in D$, $b|a$ implies b is a unit or b is an associate of a. An ideal M in a Euclidean domain D with $M \neq D$ is said to be *maximal* if for all ideals B in D, $M \subseteq B \subseteq D$ implies $B = M$ or $B = D$.

Theorem 1.11 *An element a in a Euclidean domain D is irreducible if and only if (a) is a maximal ideal in D.*

Proof. Suppose first that (a) is maximal. If $b|a$, then $(a) \subseteq (b)$. Hence, either $(b) = D$, in which case there exists $x \in D$ for which $bx = 1$ and b is a unit, or $(b) = (a)$, in which case a and b are associates. Therefore, a is irreducible. Now, suppose a is irreducible. If $(a) \subseteq (b) \subseteq D$ for some $b \in D$, then $b|a$. Hence, either b is a unit in D, in which case $(b) = D$, or a and b are associates in D, in which case $(a) = (b)$. Therefore, (a) is a maximal ideal in D.   ∎

Theorem 1.12 *An ideal M in a Euclidean domain D is maximal if and only if the quotient ring D/M is a field.*

Proof. Suppose M is a maximal ideal in D, and choose $r + M \in D/M$ such that $r + M \neq M$. Let $B = (r + M) \subseteq D/M$, and let $C = \varphi^{-1}(B)$, where φ is the ring homomorphism from D onto D/M defined by $\varphi(x) = x + M$. Since B is an ideal in D/M, by Proposition 1.8 we know that C is an ideal in D. Hence, $M \subseteq C \subseteq D$. But since M is maximal and $r + M \neq M$, then $C = D$. Therefore, $B = D/M$. Thus, there exists an element $s + M \in D/M$ for which $(r + M)(s + M) = 1 + M$, and so $r + M$ has an inverse in D/M. Hence, D/M is a field. Conversely, suppose D/M is a field, and let B be an ideal in D for which $M \subseteq B \subseteq D$. By Proposition 1.8, we know that $\varphi(B)$ is an ideal in D/M. Since the only ideals in a field are the field and $\{0\}$ (see Written Exercise 16), it follows that either $\varphi(B) = M$ or $\varphi(B) = D/M$. Hence, either $B = M$ or $B = D$, and M is maximal.   ∎

By combining the results of Theorems 1.11 and 1.12, we obtain the following theorem.

Theorem 1.13 *Suppose a is an element in a Euclidean domain D. Then the following statements are equivalent.*

1. a is irreducible in D.

2. (a) is maximal in D.

3. $D/(a)$ is a field.

1.4 Finite Fields

Finite fields play an important role in several of the applications we discuss in this book. In this section, we describe the theoretical basis of constructing finite fields. Then in Section 1.5 we demonstrate how Maple can be used to construct finite fields.

It can easily be shown (see below) that the ring $Z_p = \{0, 1, 2, \ldots, p-1\}$ for prime p is a field with the usual operations of addition and multiplication modulo p (*i.e.*, divide the result by p and take the remainder). This shows that there are finite fields of order p for every prime p. In the following discussion we show how the fields Z_p can be used to construct finite fields of order p^n for every prime p and positive integer n. A finite field of order p^n for prime p and positive integer n is sometimes called a *Galois* field, denoted $GF(p^n)$.

Let m be an irreducible element in a Euclidean domain D, and let $B = (m)$. Then by Theorem 1.13 we know that D/B is a field. If D is the ring Z of integers and $m > 0$, then m is a prime p (see Written Exercise 23). Note then that if we perform the addition and multiplication operations in D/B without including B in the notation, these operations will be exactly the addition and multiplication operations in Z_p. That is, we can view D/B as Z_p.

Now, suppose D is the integral domain $Z_p[x]$ of polynomials over Z_p for some prime p, and let $B = (f(x))$ for some irreducible polynomial $f(x)$ of degree n in D. Then again by Theorem 1.13, we know that D/B is a field. Each element in D/B is a coset of the form $g(x) + B$ for some $g(x) \in Z_p[x]$. Since $Z_p[x]$ is a Euclidean domain, then there exists $r(x) \in Z_p[x]$ for which $g(x) + B = r(x) + B$ with $r(x) = 0$ or $\deg r(x) < n$. Therefore, each element in D/B can be expressed as $r(x) + B$ for some $r(x) \in Z_p[x]$ with $r(x) = 0$ or $\deg r(x) < n$. Hence, the elements in D/B can be expressed as $r(x) + B$ for all $r(x) \in Z_p[x]$ with $r(x) = 0$ or $\deg r(x) < n$. Since a polynomial $r(x) \in Z_p[x]$ with $r(x) = 0$ or $\deg r(x) < n$ can contain up to n terms, and each of these terms can have any of p coefficients (the p elements in Z_p), then there are p^n polynomials $r(x) \in Z_p[x]$ with $r(x) = 0$ or $\deg r(x) < n$. That is, the field D/B will contain p^n distinct elements. The operations on this field are the usual operations of addition and multiplication modulo $f(x)$ (*i.e.*, divide the result by $f(x)$ and take the remainder). Because it is possible to find an irreducible polynomial of degree n over Z_p for every prime p and positive integer n, this shows that there are finite fields of order p^n for every prime p and positive integer n. It is also true that all finite fields have order p^n for some prime p and positive integer n (see Theorem 1.14).

Suppose again that $D = Z_p[x]$ for some prime p, and $B = (f(x))$ for some irreducible polynomial $f(x) \in D$. For convenience, when we write elements and perform the addition and multiplication operations in D/B, we will not include B in the notation. That is, we will write the elements $r(x) + B$ in D/B as just $r(x)$.

Example 1.9 Suppose $D = Z_3[x]$, and let $B = (f(x))$ for the irreducible polynomial $f(x) = x^2 + x + 2 \in Z_3[x]$. (Note: We can show that $f(x)$ is irreducible by verifying that $f(a) \neq 0$ for all $a \in Z_3$.) Then the field D/B will contain the $3^2 = 9$ polynomials in $Z_3[x]$ of degree less than 2. That is, $D/B = \{0, 1, 2, x, x + 1, x + 2, 2x, x + 1, 2x + 2\}$. To add elements in D/B we simply reduce the coefficients in Z_3. For example, $(2x + 1) + (2x + 2) = 4x + 3 = x$. To multiply elements in D/B we can use several methods. One method is to divide the product by $f(x)$ and take the remainder. For example, to multiply the elements $2x + 1$ and $2x + 2$ in D/B, we could form $(2x + 1)(2x + 2) = 4x^2 + 6x + 2 = x^2 + 2$. Then, dividing $x^2 + 2$ by $f(x)$, we obtain a quotient of 1 and remainder of $-x = 2x$. Hence, $(2x + 1)(2x + 2) = 2x$ in D/B. Another method for multiplying elements in D/B is to use the fact that $x^2 + x + 2 = 0$ in D/B. Therefore, $x^2 = -x - 2 = 2x + 1$ in D/B. The identity $x^2 = 2x + 1$ can then be used to reduce powers of x in D/B. For example, we can also compute the product of the elements $2x + 1$ and $2x + 2$ in D/B by forming $(2x + 1)(2x + 2) = 4x^2 + 6x + 2 = x^2 + 2 = (2x + 1) + 2 = 2x$. A third method for multiplying elements in D/B will be described in general next and then illustrated in Example 1.10. ∎

A fundamental fact regarding finite fields is that the nonzero elements in every finite field form a cyclic multiplicative group (see Theorem 1.15). Suppose $D = Z_p[x]$ for some prime p, and $B = (f(x))$ for some irreducible polynomial $f(x) \in D$. For the field $F = D/B$, if x is a cyclic generator for F^*, then $f(x)$ is said to be *primitive*. Hence, if $f(x)$ is primitive, then all nonzero elements in F can be generated by constructing powers of x modulo $f(x)$. This is useful because it allows products of elements in F to be formed by converting the elements to their representations as powers of x, multiplying the powers of x, and then converting the result back to an element in F. This is illustrated in the following example.

Example 1.10 Consider the field D/B in Example 1.9. In this field we can use the identity $x^2 = 2x + 1$ to construct the elements that correspond to powers of x. For example, we can construct the field element that corresponds to x^3 as follows.

$$x^3 = xx^2 = x(2x + 1) = 2x^2 + x = 2(2x + 1) + x = 5x + 2 = 2x + 2$$

Hence, $x^3 = 2x + 2$ in D/B. And we can construct the field element that corresponds to x^4 as follows.

$$x^4 = xx^3 = x(2x + 2) = 2x^2 + 2x = 2(2x + 1) + 2x = 6x + 2 = 2$$

Therefore, $x^4 = 2$ in D/B. The field elements that correspond to subsequent powers of x can be constructed similarly. We list the field elements that correspond to the first 8 powers of x in the following table.

Power	Field Element
x^1	x
x^2	$2x + 1$
x^3	$2x + 2$
x^4	2
x^5	$2x$
x^6	$x + 2$
x^7	$x + 1$
x^8	1

The only element in D/B not listed in this table is 0. Since all nonzero elements in D/B are in the cyclic group generated by x, then $f(x) = x^2 + x + 2$ is primitive in $Z_3[x]$.

The preceding table is useful for computing products in D/B. For example, we can form the product of the elements $2x + 1$ and $2x + 2$ in D/B as follows.

$$(2x + 1)(2x + 2) = x^2 x^3 = x^5 = 2x$$

Note that this matches the product obtained in Example 1.9. And we can form the product of the elements $2x$ and $x + 2$ in D/B as follows.

$$(2x)(x + 2) = x^5 x^6 = x^{11} = x^8 x^3 = 1x^3 = 2x + 2$$

Other products in D/B can be formed similarly.   ■

Example 1.11 Suppose $D = Z_3[x]$, and let $B = (f(x))$ for the polynomial $f(x) = x^2 + 1 \in Z_3[x]$. Since $f(x)$ is irreducible in $Z_3[x]$, then D/B is a field of order $3^2 = 9$ (with the same elements as the field in Example 1.9). However, note that $x^2 = -1 = 2$ in D/B, and thus $x^4 = 4 = 1$ in D/B. Hence, computing powers of x will not generate all 8 nonzero elements in D/B. Therefore, $f(x) = x^2 + 1$ is not primitive in $Z_3[x]$, and we cannot compute all possible products in D/B using the method illustrated in Example 1.10. However, we can still compute all possible products in D/B using the methods illustrated in Example 1.9.   ■

We close this section by proving two fundamental results we have mentioned regarding finite fields.

Theorem 1.14 *Suppose F is a finite field. Then $|F| = p^n$ for some prime p and positive integer n.*

Proof. Let H be the additive subgroup of F generated by 1. Suppose $|H| = mn$ for some positive integers m, n with $m \neq 1$ and $n \neq 1$. Then $0 = (mn)1 = (m1)(n1)$. But since $m1 \neq 0$ and $n1 \neq 0$, this contradicts the fact that F is a field. Hence, $|H| = p$ for some prime p. That is, $H = Z_p$ for some prime p. The field F can then be viewed as a vector space over H with scalar multiplication given by the field multiplication, so F has a basis with a finite number of elements, say n. The order of F is the number p^n of linear combinations of these basis elements over Z_p.   ∎

Theorem 1.15 *Let F be a finite field. Then F^* is a cyclic multiplicative group.*

Proof. Clearly, F^* is an abelian multiplicative group. To show that F^* is cyclic, we use the first of the well-known Sylow Theorems, which states that for a finite group G of order n, if p^k divides n for some prime p and positive integer k, then G contains a subgroup of order p^k. Suppose $|F^*|$ has prime factorization $p_1^{n_1} p_2^{n_2} \cdots p_t^{n_t}$, and let S_i be subgroups of order $p_i^{n_i}$ in F^* for each $i = 1, 2, \ldots, t$. Let $k_i = p_i^{n_i - 1}$ for each $i = 1, 2, \ldots, t$. Then, if S_i is not cyclic for some i, it follows that $a^{k_i} = 1$ for all $a \in S_i$. Hence, $f(x) = x^{k_i} - 1$ has $p_i^{n_i}$ roots in F, a contradiction. Thus, each S_i must have a cyclic generator a_i. Let $b = a_1 a_2 \cdots a_t$. Since b has order $|F^*|$, then b is a cyclic generator for F^*.   ∎

1.5   Finite Fields with Maple

In this section, we show how Maple can be used to construct the nonzero elements in a finite field $Z_p[x]/(f(x))$ for prime p and primitive polynomial $f(x) \in Z_p[x]$ as powers of x. We consider the field in Example 1.10.

We begin by defining the polynomial $f(x) = x^2 + x + 2 \in Z_3[x]$ used to construct the field elements.

```
> f := x -> x^2 + x + 2;
```

$$f := x \rightarrow x^2 + x + 2$$

We can use the Maple **Irreduc** function to verify that $f(x)$ is irreducible in $Z_3[x]$. The following command will return *true* if $f(x)$ is irreducible modulo 3, and *false* if not.

```
> Irreduc(f(x)) mod 3;
```

$$true$$

Hence, $f(x)$ is irreducible in $Z_3[x]$, and $Z_3[x]/(f(x))$ is a field. However, in order for us to be able to construct all of the nonzero elements in this field by computing powers of x, $f(x)$ must also be primitive. We can use the Maple **Primitive** function to verify that $f(x)$ is primitive in $Z_3[x]$. The following command will return *true* if $f(x)$ is primitive modulo 3, and *false* if not.

```
> Primitive(f(x)) mod 3;
```

$$true$$

Therefore, $f(x)$ is primitive in $Z_3[x]$.

To construct elements in $Z_3[x]/(f(x))$ as powers of x, we can use the Maple **Powmod** function. For example, the following command returns x^6 modulo $f(x)$.

```
> Powmod(x, 6, f(x), x) mod 3;
```

$$x + 2$$

In the preceding command, the polynomial x given by the first parameter is raised to the power 6 given by the second parameter, with the output displayed after the result is reduced modulo the third parameter $f(x)$ (defined over the specified modulus 3). The fourth parameter is the variable used in the first and third parameters.

We will now use a Maple **for** loop to construct and display all of the 8 nonzero elements in $Z_3[x]/(f(x))$ and corresponding powers of x. In the following commands, we store the results returned by **Powmod** for each of the first 8 powers of x in the variable **temp** and display these results using the Maple **print** command. Note where we use colons and semicolons in this loop. Note also that we use back ticks " ' " in the **print** statement.

```
> for i from 1 to 8 do
>       temp := Powmod(x, i, f(x), x) mod 3:
>       print(x^i, '    Field Element:  ', temp);
> od:
```

$$x, \quad \textit{Field Element:} \quad , x$$

$$x^2, \quad \textit{Field Element:} \quad , 2x + 1$$

$$x^3, \quad \textit{Field Element:} \quad , 2x + 2$$

$$x^4, \quad \textit{Field Element:} \quad , 2$$

$$x^5, \quad \textit{Field Element:} \quad , 2x$$

$$x^6, \quad \textit{Field Element:} \quad , x + 2$$

$$x^7, \quad \textit{Field Element:} \quad , x + 1$$

$$x^8, \quad \textit{Field Element:} \quad , 1$$

Note that these results match those listed in Example 1.10 for the nonzero elements in $Z_3[x]/(f(x))$.

1.6   The Euclidean Algorithm

Let a and b be nonzero elements in a Euclidean domain D, and consider an element $d \in D$ for which $d|a$ and $d|b$. Suppose that for all $x \in D$, if $x|a$ and $x|b$, then $x|d$. Then d is called a *greatest common divisor* of a and b. We will use the notation $d = (a, b)$ to represent this.

Greatest common divisors do not always exist for two elements in a general ring. But as we will show in Theorem 1.16, greatest common divisors do always exist for two elements in a Euclidean domain. As they are defined above, there is not a unique greatest common divisor of two elements in a Euclidean domain. For example, in the ring Z of integers, both 1 and -1 are greatest common divisors of any two distinct primes. However, it can be shown very easily that if both d_1 and d_2 are greatest common divisors of two elements in a Euclidean domain D, then d_1 and d_2 are associates in D (see Written Exercise 30).

Theorem 1.16 *Let a and b be nonzero elements in a Euclidean domain D. Then there exists a greatest common divisor d of a and b that can be expressed as $d = au + bv$ for some $u, v \in D$.*

Proof.   Let B be an ideal in D of smallest order that contains both a and b. It can easily be shown that $B = \{ar + bs \mid r, s \in D\}$ (see Written Exercise 31). Since D is a Euclidean domain, by Theorem 1.9 we know that D is a principal ideal domain. Hence, $B = (d)$ for some $d \in D$. Since d generates B, and $a, b \in B$, then $d|a$ and $d|b$. And since $d \in B = \{ar + bs \mid r, s \in D\}$, then $d = au + bv$ for some $u, v \in D$. Now, if $x|a$ and $x|b$ for some $x \in D$, then $a = xr$ and $b = xs$ for some $r, s \in D$. Therefore, $d = au + bv = xru + xsv = x(ru + sv)$, and $x|d$. ∎

When considering only certain specific rings, it is often convenient to place restrictions on greatest common divisors to make them unique. For example, for elements a and b in the ring Z of integers, there is a unique positive greatest common divisor of a and b. And for elements a and b in the ring $F[x]$ of polynomials over a field F, there is a unique greatest common divisor of a and b that is monic (*i.e.*, that has a leading coefficient of 1). Since these are the only rings we will use extensively here, for the remainder of this book we will assume greatest common divisors are defined uniquely with these restrictions. We should note that even though the greatest common divisor (a, b) of two integers or polynomials a and b is uniquely defined with these restrictions, the u and v that yield $(a, b) = au + bv$ need not be unique.

In several of the applications in this book we will need to determine not only the greatest common divisor (a, b) of two integers or polynomials a and b, but also u and v that yield $(a, b) = au + bv$. We will use the Euclidean algorithm to do this. We describe this algorithm next.

Let a and b be nonzero elements in a Euclidean domain D, and let N be the set of nonnegative integers. Since D is a Euclidean domain, then there is a mapping $\delta : D^* \rightarrow N$ for which we can find $q_1, r_1 \in D$ with $a = bq_1 + r_1$ and $r_1 = 0$ or $\delta(r_1) < \delta(b)$. Suppose $\delta(r_1) < \delta(b)$. Then we can find $q_2, r_2 \in D$ with $b = r_1 q_2 + r_2$ and $r_2 = 0$ or $\delta(r_2) < \delta(r_1)$. Suppose $\delta(r_2) < \delta(r_1)$. Then we can find $q_3, r_3 \in D$ with $r_1 = r_2 q_3 + r_3$ and $r_3 = 0$ or $\delta(r_3) < \delta(r_2)$. We continue this process until the first time $r_i = 0$ (which is guaranteed to happen eventually since the $\delta(r_i)$ form a strictly decreasing sequence of nonnegative integers). That is, we construct all q_i, r_i for the following equations.

$$
\begin{aligned}
a &= bq_1 + r_1 & \{\delta(r_1) < \delta(b)\} \\
b &= r_1 q_2 + r_2 & \{\delta(r_2) < \delta(r_1)\} \\
r_1 &= r_2 q_3 + r_3 & \{\delta(r_3) < \delta(r_2)\} \\
&\vdots \\
r_{n-2} &= r_{n-1} q_n + r_n & \{\delta(r_n) < \delta(r_{n-1})\} \\
r_{n-1} &= r_n q_{n+1} + 0
\end{aligned}
$$

By working up this list of equations we can see that r_n divides both a and b. By working down the list we can see that any $x \in D$ that divides both a and b must also divide r_n. Hence, $(a, b) = r_n$. This technique for determining (a, b) is called the *Euclidean algorithm*.

We have now shown a technique for determining the greatest common divisor (a, b) of two integers or polynomials a and b. We must still show a technique for finding u and v that yield $(a, b) = au + bv$. To do this, we consider the following table constructed using the q_i, r_i from the preceding list of equations, and u_i, v_i we describe below. We will call this table a *Euclidean algorithm table.*

Row	Q	R	U	V
-1	$-$	$r_{-1} = a$	$u_{-1} = 1$	$v_{-1} = 0$
0	$-$	$r_0 = b$	$u_0 = 0$	$v_0 = 1$
1	q_1	r_1	u_1	v_1
2	q_2	r_2	u_2	v_2
$\vdots$	$\vdots$	$\vdots$	$\vdots$	$\vdots$
n	q_n	r_n	u_n	v_n

The entries in each row $i = 1, 2, \ldots, n$ of this table are constructed as follows. The q_i, r_i are from the ith equation

$$r_{i-2} \quad = \quad r_{i-1}q_i + r_i \tag{1.1}$$

in the preceding list of equations. Note that if we solve for r_i in (1.1), we obtain the following equation.

$$r_i \quad = \quad r_{i-2} - r_{i-1}q_i \tag{1.2}$$

We then construct u_i, v_i by following this pattern for constructing r_i from q_i. Specifically, we construct u_i, v_i from q_i as follows.

$$u_i \quad = \quad u_{i-2} - u_{i-1}q_i \tag{1.3}$$
$$v_i \quad = \quad v_{i-2} - v_{i-1}q_i \tag{1.4}$$

Many useful relations exist between the entries in a Euclidean algorithm table. For example, the following equation is true for all rows i.

$$r_i \quad = \quad au_i + bv_i \tag{1.5}$$

Clearly, this equation is true for rows $i = -1$ and 0. To see that it is true for all subsequent rows, assume it is true for all rows i through $k - 1$. Then, using (1.2), (1.3), and (1.4), it follows that

$$
\begin{aligned}
r_k \quad &= \quad r_{k-2} - r_{k-1}q_k \\
&= \quad (au_{k-2} + bv_{k-2}) - (au_{k-1} + bv_{k-1})q_k \\
&= \quad a(u_{k-2} - u_{k-1}q_k) + b(v_{k-2} - v_{k-1}q_k) \\
&= \quad au_k + bv_k.
\end{aligned}
$$

Specifically, $r_n = a u_n + b v_n$. But recall, we have stated that $r_n = (a, b)$. Hence, for $u = u_n$ and $v = v_n$, we have $(a, b) = au + bv$.

Another useful relation between the entries in a Euclidean algorithm table is the following equation for all $i = -1, 0, 1, 2, \ldots, n-1$.

$$r_i u_{i+1} - u_i r_{i+1} = (-1)^i b \qquad (1.6)$$

Note first that this equation is clearly true for $i = -1$. To see that it is true for all subsequent i, assume it is true for $i = k-1$. Then, using (1.2), (1.3), and the fact that adding a multiple of a row of a matrix to another row in the matrix does not change the determinant of the matrix, it follows that

$$
\begin{aligned}
r_k u_{k+1} - u_k r_{k+1} &= \begin{vmatrix} r_k & u_k \\ r_{k+1} & u_{k+1} \end{vmatrix} \\
&= \begin{vmatrix} r_k & u_k \\ r_{k-1} - r_k q_{k+1} & u_{k-1} - u_k q_{k+1} \end{vmatrix} \\
&= \begin{vmatrix} r_k & u_k \\ r_{k-1} & u_{k-1} \end{vmatrix} \\
&= r_k u_{k-1} - u_k r_{k-1} \\
&= -(r_{k-1} u_k - u_{k-1} r_k) \\
&= -(-1)^{k-1} b \\
&= (-1)^k b.
\end{aligned}
$$

Two additional relations that exist between the entries in a Euclidean algorithm table are the following equations for all $i = -1, 0, 1, 2, \ldots, n-1$.

$$r_i v_{i+1} - v_i r_{i+1} = (-1)^{i+1} a \qquad (1.7)$$

$$u_i v_{i+1} - u_{i+1} v_i = (-1)^{i+1} \qquad (1.8)$$

These equations can be verified in a manner similar to the verification of (1.6) given above (see Written Exercises 32 and 33).

We close this section with two examples in which we use the Euclidean algorithm to find (a, b), and a Euclidean algorithm table to find u and v such that $(a, b) = au + bv$.

Example 1.12 In this example, we consider $a = 81$ and $b = 64$ in Z. To use the Euclidean algorithm to find (a, b), we form the following equations.

$$
\begin{aligned}
81 &= 64 \cdot 1 + 17 \\
64 &= 17 \cdot 3 + 13 \\
17 &= 13 \cdot 1 + 4 \\
13 &= 4 \cdot 3 + 1 \\
4 &= 1 \cdot 4 + 0
\end{aligned}
$$

Hence, $(81, 64) = 1$. It can easily be verified that these equations yield the following Euclidean algorithm table.

Row	Q	R	U	V
−1	−	81	1	0
0	−	64	0	1
1	1	17	1	−1
2	3	13	−3	4
3	1	4	4	−5
4	3	1	−15	19

Thus, $u = -15$ and $v = 19$ satisfy $(81, 64) = 81u + 64v$. ■

Example 1.13 In this example, we consider $a = x^6 + x^2 + x$ and $b = x^4 + x^2 + x$ in $Z_2[x]$. To use the Euclidean algorithm to find (a, b), we form the following equations.

$$
\begin{aligned}
a &= b(x^2 + 1) + x^3 \\
b &= x^3(x) + (x^2 + x) \\
x^3 &= (x^2 + x)(x + 1) + x \\
x^2 + x &= x(x + 1) + 0
\end{aligned}
$$

Therefore, $(a, b) = x$. The u_i and v_i for the resulting Euclidean algorithm table are constructed as follows (with all coefficients expressed in Z_2).

$$
\begin{aligned}
u_1 &= u_{-1} - u_0 q_1 = 1 - 0(x^2 + 1) & &= 1 \\
v_1 &= v_{-1} - v_0 q_1 = 0 - 1(x^2 + 1) & &= x^2 + 1 \\
u_2 &= u_0 - u_1 q_2 = 0 - 1x & &= x \\
v_2 &= v_0 - v_1 q_2 = 1 - (x^2 + 1)x & &= x^3 + x + 1 \\
u_3 &= u_1 - u_2 q_3 = 1 - x(x + 1) & &= x^2 + x + 1 \\
v_3 &= v_1 - v_2 q_3 = (x^2 + 1) - (x^3 + x + 1)(x + 1) & &= x^4 + x^3
\end{aligned}
$$

Thus, the Euclidean algorithm table is the following.

Row	Q	R	U	V
−1	−	$x^6 + x^2 + x$	1	0
0	−	$x^4 + x^2 + x$	0	1
1	$x^2 + 1$	x^3	1	$x^2 + 1$
2	x	$x^2 + x$	x	$x^3 + x + 1$
3	$x + 1$	x	$x^2 + x + 1$	$x^4 + x^3$

Hence, $u = x^2 + x + 1$ and $v = x^4 + x^3$ satisfy $(a, b) = au + bv$. ■

Written Exercises

1. Let $A(S)$ be the set of bijections on a set S, and let $\circ$ be the composition operation on $A(S)$. Show that $(A(S), \circ)$ is a group.

2. Show that $|S_n| = n!$ for the symmetric group S_n.

3. Consider the following elements in S_6:

$$\alpha : \begin{pmatrix} 1 & 2 & 3 & 4 & 5 & 6 \\ 4 & 3 & 6 & 2 & 1 & 5 \end{pmatrix}$$

$$\beta : \begin{pmatrix} 1 & 2 & 3 & 4 & 5 & 6 \\ 4 & 1 & 6 & 2 & 3 & 5 \end{pmatrix}$$

$$\gamma : \begin{pmatrix} 1 & 2 & 3 & 4 & 5 & 6 \\ 4 & 1 & 6 & 2 & 5 & 3 \end{pmatrix}$$

   (a) Find the elements $\alpha \circ \beta$ and $\beta \circ \gamma$ in S_6, where $\circ$ represents the composition operation.

   (b) Express α, β, and γ as a cycle or product of disjoint cycles.

   (c) Is $\alpha \circ \gamma$ even or odd?

   (d) Find the inverses of α, β, and γ.

   (e) Express α, β, and γ as a product of transpositions.

4. Find the elements in the alternating group A_4.

5. Find the elements in the dihedral group D_3.

6. Find the elements in $A_5 \cap D_5$.

7. Find the distinct left cosets of A_4 in S_4.

8. Show that A_3 is cyclic.

9. Find the order of the following elements.

   (a) The $144°$ rotation in D_5.

   (b) The $144°$ rotation in D_{10}.

   (c) Reflection across horizontal in D_{10}.

   (d) The element α in Written Exercise 3.

   (e) The element $(123)(45)(67)$ in A_7.

10. Show that if a group G is cyclic, then G is abelian.

11. Show that if H is a subgroup of a cyclic group, then H is cyclic.

12. Show that if H is a subgroup of a cyclic group G, then G/H is cyclic.

13. Find the kernel of the homomorphisms in Examples 1.7 and 1.8.

14. Let G and H be groups, and suppose $\varphi : G \to H$ is a homomorphism. Show that Ker φ is a normal subgroup of G.

15. Show that A_n is a normal subgroup of S_n.

16. Show that the only ideals in a field F are F and $\{0\}$.

17. Let a be an element in a field F. Define the mapping $\varphi : F[x] \to F$ by $\varphi(f(x)) = f(a)$. Show that φ is a ring homomorphism, and find Ker φ.

18. Prove Proposition 1.8.

19. Show that the ring $F[x]$ of polynomials over a field F is a Euclidean domain with the function $\delta(f(x)) = \deg f(x)$.

20. Is it true that all ideals in the ring $F[x]$ of polynomials over a field F are principal ideals? State how you know.

21. Show that the ring Z of integers is a Euclidean domain with the function $\delta(n) = |n|$.

22. Prove Proposition 1.10.

23. Find all irreducible elements in the ring Z of integers.

24. Perform the following calculations.

   (a) $(x + 2) + (2x + 2)$ in the field D/B in Examples 1.9 and 1.10.

   (b) $(x + 2)(2x + 2)$ in the field D/B in Examples 1.9 and 1.10.

   (c) $(x + 2) + (2x + 2)$ in the field D/B in Example 1.11.

   (d) $(x + 2)(2x + 2)$ in the field D/B in Example 1.11.

25. Let $f(x) = x^2 + x + 2$.

   (a) Show that $f(x)$ is primitive in $Z_3[x]$ by constructing the field elements that correspond to powers of x in $Z_3[x]/(f(x))$.

   (b) Show that $f(x)$ is primitive in $Z_5[x]$ by constructing the field elements that correspond to powers of x in $Z_5[x]/(f(x))$.

   (c) Show that $f(x)$ is not primitive in $Z_{11}[x]$ by showing that $f(x)$ is not irreducible in $Z_{11}[x]$.

26. Show that $f(x) = x^3 + x + 1$ is primitive in $Z_2[x]$ by constructing the field elements that correspond to powers of x in $Z_2[x]/(f(x))$.

27. Show that $f(x) = x^3 + x^2 + 1$ is primitive in $Z_2[x]$ by constructing the field elements that correspond to powers of x in $Z_2[x]/(f(x))$.

28. Show that $f(x) = x^4 + x + 1$ is primitive in $Z_2[x]$ by constructing the field elements that correspond to powers of x in $Z_2[x]/(f(x))$.

29. Let $f(x) = x^4 + x^3 + x^2 + x + 1$, $g(x) = x^4 + x^3 + x^2 + 1$, and $h(x) = x^4 + x^3 + 1$. In $Z_2[x]$, one of the polynomials $f(x)$, $g(x)$, and $h(x)$ is primitive, one is irreducible but not primitive, and one is not irreducible. Which is which? Explain how you know. For the polynomial that is irreducible but not primitive, find the multiplicative order of x.

30. Show that if d_1 and d_2 are greatest common divisors of two elements in an integral domain D, then d_1 and d_2 are associates in D.

31. Let a and b be elements in an integral domain D, and let B be an ideal in D of smallest order that contains both a and b. Show that $B = \{ar + bs \mid r, s \in D\}$.

32. Verify Equation (1.7).

33. Verify Equation (1.8).

34. Use the Euclidean algorithm to find $(2272, 716)$, and use a Euclidean algorithm table to find u and v such that $(2272, 716) = 2272u + 716v$.

35. Let $a = x^5 + x^4 + x^3 + x^2$ and $b = x^4 + x^3 + x + 1$ in $Z_2[x]$. Use the Euclidean algorithm to find (a, b), and use a Euclidean algorithm table to find u and v such that $(a, b) = au + bv$.

Maple Exercises

1. Find a primitive polynomial of degree 4 in $Z_3[x]$, and use this polynomial to construct the nonzero elements in a finite field.

2. Find a primitive polynomial of degree 2 in $Z_{11}[x]$, and use this polynomial to construct the nonzero elements in a finite field.

3. Construct the nonzero elements in a finite field of order 128.

4. Construct the nonzero elements in a finite field of order 127.

Chapter 2

Block Designs

Suppose a magazine editor wishes to compare seven cars by evaluating the responses of seven consumers to a series of questions regarding topics such as handling and comfort. The most obvious way for the editor to obtain a valid comparison of the cars would be to have each of the consumers test each of the cars. However, for various reasons such as time or monetary constraints, it may not be feasible to have each of the consumers test each car. The most convenient way to obtain a comparison of the cars would be to have each of the consumers test just one of the cars. But this might not yield a valid comparison of the cars due to potential differences among the consumers. In this chapter, we discuss some techniques the editor could use to ensure a testing scheme that is both fair and reasonable.

2.1   General Properties of Block Designs

Let $B_1, \ldots, B_b$ be subsets of a set $S = \{a_1, \ldots, a_v\}$. We will call the elements a_i *objects* and the subsets B_j *blocks*. This collection of objects and blocks is called a *balanced incomplete block design* if it satisfies the following conditions:

1. Each block contains the same number of objects.

2. Each object is contained in the same number of blocks.

3. Each pair of objects appears together in the same number of blocks.

For convenience, we will refer to balanced incomplete block designs as just *block designs*. A block design is described by parameters (v, b, r, k, λ) if it has v objects and b blocks, each object is contained in r blocks, each block contains k objects, and each pair of objects appears together in λ blocks.

In all of the (v, b, r, k, λ) block designs we consider in this book, we will assume $k < v$ and $\lambda > 0$. These restrictions are harmless, for clearly $k \leq v$, and $k = v$ corresponds to the case when each block contains all of the objects. With regard to the example in the introduction to this chapter, this represents the possibly infeasible case when each of the consumers (represented by the blocks) tests each of the cars (represented by the objects). Also, clearly $\lambda \geq 0$, and $\lambda = 0$ corresponds to the case when each block contains only one object. With regard to the example in the introduction to this chapter, this represents the possibly invalid case when each of the consumers tests just one of the cars.

Example 2.1 Suppose a magazine editor wishes to obtain a fair and reasonable comparison of seven cars by evaluating the opinions of seven consumers. If we represent the cars by the elements in the set $S = \{1, 2, 3, 4, 5, 6, 7\}$, then each consumer can be represented by a block containing the cars to be tested by that consumer. For example, the subsets $\{1, 2, 4\}, \{2, 3, 5\}, \{3, 4, 6\}, \{4, 5, 7\}, \{5, 6, 1\}, \{6, 7, 2\}$, and $\{7, 1, 3\}$ of S are the blocks in a $(7, 7, 3, 3, 1)$ block design, indicating that the first consumer should test cars 1, 2, and 4, the second consumer should test cars 2, 3, and 5, and so forth. Note that in this block design, each car is tested three times, each consumer tests three cars, and each pair of cars is tested by the same consumer once. Therefore, this design yields a valid comparison of the cars while requiring only 21 total tests (versus 49 tests if each consumer tests each car). ∎

In this chapter we discuss several techniques for constructing block designs, including one that yields the design in Example 2.1. Before discussing these techniques, we first mention some general properties of block designs.

Theorem 2.1 *The parameters in a* (v, b, r, k, λ) *block design satisfy the equations* $vr = bk$ *and* $(v - 1)\lambda = r(k - 1)$.

Proof. To show that the equation $vr = bk$ holds, we consider the set $T = \{(a, B) \mid a \text{ is an object in block } B\}$, and count $|T|$ in two ways. First, the design has v objects that each appear in r blocks. Hence, $|T| = vr$. But the design also has b blocks that each contain k objects. Hence, $|T| = bk$. Thus, $vr = bk$. To show that $(v - 1)\lambda = r(k - 1)$, we choose an object a_0

in the design. Then for $U = \{(x, B) \mid x$ is an object with a_0 in block $B\}$, we count $|U|$ in two ways. First, there are $v - 1$ objects in the design that each appear in λ blocks with a_0, so $|U| = (v - 1)\lambda$. But there are also r blocks in the design that each contain a_0 and $k - 1$ other objects. Hence, $|U| = r(k - 1)$. Thus, $(v - 1)\lambda = r(k - 1)$. ∎

For a block design with objects $a_1, \ldots, a_v$ and blocks $B_1, \ldots, B_b$, let $A = (a_{ij})$ be the $v \times b$ matrix for which $a_{ij} = 1$ if $a_i \in B_j$, and $a_{ij} = 0$ if $a_i \notin B_j$. Then A is called an *incidence matrix* for the design.

Example 2.2 The following is the incidence matrix for the block design in Example 2.1 with objects and blocks taken in order of appearance.

$$\begin{bmatrix}
1 & 0 & 0 & 0 & 1 & 0 & 1 \\
1 & 1 & 0 & 0 & 0 & 1 & 0 \\
0 & 1 & 1 & 0 & 0 & 0 & 1 \\
1 & 0 & 1 & 1 & 0 & 0 & 0 \\
0 & 1 & 0 & 1 & 1 & 0 & 0 \\
0 & 0 & 1 & 0 & 1 & 1 & 0 \\
0 & 0 & 0 & 1 & 0 & 1 & 1
\end{bmatrix}$$

∎

In this chapter we use incidence matrices for two purposes. In Section 2.2 we use them to construct block designs. In this section we use them to prove some general results about block designs.

Let A be an incidence matrix for a (v, b, r, k, λ) block design. Note that the dot product of any row i of A with itself will be equal to the number r of blocks in the design that contain a_i. Note also that the dot product of any two distinct rows i and j of A will be equal to the number λ of blocks in the design that contain both a_i and a_j. Since the matrix AA^t can be viewed as containing the dot product of every row of A with itself and all other rows of A, then

$$AA^t = \begin{bmatrix}
r & \lambda & \ldots & \lambda \\
\lambda & r & \ldots & \lambda \\
\vdots & \vdots & & \vdots \\
\lambda & \lambda & \ldots & r
\end{bmatrix} = (r - \lambda)I + \lambda J,$$

where I is the $v \times v$ identity matrix, and J is the $v \times v$ matrix of all ones.

Lemma 2.2 *Let B be a $v \times v$ matrix such that $B = (r - \lambda)I + \lambda J$, where I is the $v \times v$ identity matrix and J is the $v \times v$ matrix of all ones. Then* $\det B = (r - \lambda)^{(v-1)}(r + (v - 1)\lambda)$.

Proof. Note first that B must have the following form.

$$B = \begin{bmatrix} r & \lambda & \cdots & \lambda \\ \lambda & r & \cdots & \lambda \\ \vdots & \vdots & & \vdots \\ \lambda & \lambda & \cdots & r \end{bmatrix}$$

Subtracting the first column of B from each of the remaining columns of B yields the following.

$$B_1 = \begin{bmatrix} r & \lambda - r & \cdots & \lambda - r \\ \lambda & r - \lambda & \cdots & 0 \\ \vdots & \vdots & & \vdots \\ \lambda & 0 & \cdots & r - \lambda \end{bmatrix}$$

Then, adding each row of B_1 except the first to the first row of B_1 yields the following.

$$B_2 = \begin{bmatrix} r + (v - 1)\lambda & 0 & \cdots & 0 \\ \lambda & r - \lambda & \cdots & 0 \\ \vdots & \vdots & & \vdots \\ \lambda & 0 & \cdots & r - \lambda \end{bmatrix}$$

Since B_2 is triangular, $\det B_2$ is equal to the product of the diagonal entries of B_2. Hence, $\det B_2 = (r - \lambda)^{(v-1)}(r + (v - 1)\lambda)$. But $\det B = \det B_2$. Thus, $\det B = (r - \lambda)^{(v-1)}(r + (v - 1)\lambda)$.   ∎

Theorem 2.3 *The parameters in a* (v, b, r, k, λ) *block design satisfy the inequalities* $v \le b$ *and* $k \le r$.

Proof. Let A be an incidence matrix for the design. Since $k < v$, Theorem 2.1 implies $\lambda < r$. Then by Lemma 2.2, we know $\det AA^t \ne 0$. Since the rank of a product is at most the minimum rank of the factors, it follows that rank $A \ge$ rank $AA^t = v$. Hence, since A is of size $v \times b$, we know that $v \le b$. And then by Theorem 2.1 we know that $k \le r$.   ∎

A block design is said to be *symmetric* if it has the same number of objects and blocks. That is, a (v, b, r, k, λ) block design is symmetric if $b = v$, which by Theorem 2.1 implies $k = r$. The block design in Example 2.1 is symmetric.

Theorem 2.4 *In a* (v, v, r, r, λ) *block design, each distinct pair of blocks contains* λ *objects in common.*

Proof. Let A be an incidence matrix for the design. By Lemma 2.2 we know that A must be nonsingular. Also, for the $v \times v$ matrix J of all ones, it follows that $AJ = JA$ since each entry in both products will be r. Now, since $AA^t = (r - \lambda)I + \lambda J$ for the $v \times v$ identity matrix I, and $AJ = JA$, then
$$AA^t A = ((r - \lambda)I + \lambda J)A = A((r - \lambda)I + \lambda J) = AAA^t.$$
Since A is nonsingular, it can be canceled from the left of both sides of the equation $AA^t A = AAA^t$, leaving $A^t A = AA^t = (r - \lambda)I + \lambda J$. Thus, the dot product of any two distinct columns of A (the off-diagonal entries of $A^t A$) will be equal to λ. Hence, each distinct pair of blocks in the design will contain λ objects in common.   ∎

Theorem 2.4 states that in a symmetric block design, the number of objects contained in common in each pair of blocks will be equal to the number of blocks that contain each pair of objects. Thus, in the block design in Example 2.1, each pair of consumers will test the same car once.

2.2 Hadamard Matrices

In this section we show how Hadamard matrices can be used to construct block designs. An $n \times n$ matrix H is called a *Hadamard matrix* if the entries in H are all 1 or -1, and $HH^t = nI$ for the $n \times n$ identity matrix I.

For an $n \times n$ Hadamard matrix H, since $\frac{1}{n}H^t = H^{-1}$, then $H^t H = nI$. Since $HH^t = H^t H = nI$, we see that the dot product of any row or column of H with itself will be equal to n, and the dot product of any two distinct rows or columns of H will be equal to 0. Thus, changing the sign of each entry in a row or column of H will yield another Hadamard matrix. A Hadamard matrix H is said to be *normalized* if the first row and column of H contain only positive ones. Therefore, every Hadamard matrix can be converted into a normalized Hadamard matrix by changing the signs of the entries in the necessary rows and columns. Because the first row and column of a normalized Hadamard matrix H contain only positive ones, all other rows and columns of H must contain the same number of positive and negative ones. Thus, for a Hadamard matrix H of order n, if $n > 1$, then n must be even. In fact, if $n > 2$, then n must be a multiple of 4, since for $H = (h_{ij})$,

$$\sum_j (h_{1j} + h_{2j})(h_{1j} + h_{3j}) = \sum_j h_{1j}^2 = n,$$

and $(h_{1j} + h_{2j})(h_{1j} + h_{3j}) = 0$ or 4 for each j.

The only normalized Hadamard matrices of orders one and two (*i.e.*, of sizes 1×1 and 2×2) are $H_1 = \begin{bmatrix} 1 \end{bmatrix}$ and $H_2 = \begin{bmatrix} 1 & 1 \\ 1 & -1 \end{bmatrix}$. Also, $H_4 = \begin{bmatrix} H_2 & H_2 \\ H_2 & -H_2 \end{bmatrix}$ is a normalized Hadamard matrix of order four. This construction of H_4 from H_2 generalizes. Specifically, if H is a (normalized) Hadamard matrix, then $\begin{bmatrix} H & H \\ H & -H \end{bmatrix}$ is also a (normalized) Hadamard matrix (see Written Exercise 6). This shows that there are Hadamard matrices of order 2^n for every positive integer n.

We are interested in Hadamard matrices because they provide us with a method for constructing block designs. For a normalized Hadamard matrix H of order $4t \geq 8$, if we delete the first row and column from H, and change all negative ones in H to zeros, the resulting matrix will be an incidence matrix for a $(4t - 1, 4t - 1, 2t - 1, 2t - 1, t - 1)$ block design. We state this as the following theorem.

Theorem 2.5 *Let H be a normalized Hadamard matrix of order $4t \geq 8$. If the first row and column of H are deleted, and all negative ones in H are changed to zeros, the resulting matrix will be an incidence matrix for a $(4t - 1, 4t - 1, 2t - 1, 2t - 1, t - 1)$ block design.*

Proof. Delete the first row and column from H, change all negative ones in H to zeros, and call the resulting matrix A. Each row and column of H except the first will contain $2t$ ones. Therefore, each row and column of A will contain $2t - 1$ ones. Hence, the dot product of any row or column of A with itself will be equal to $k = 2t - 1$. Furthermore, in any pair of distinct rows of H excluding the first, there will be $2t$ positions in which the rows differ, t positions in which the rows both have a 1, and t positions in which the rows both have a -1. Thus, in the corresponding pair of rows of A, there will be $t - 1$ positions in which the rows both have a 1, so the dot product of any two distinct rows of A will be equal to $\lambda = t - 1$. Therefore, $AA^t = (k - \lambda)I + \lambda J$, where I is the $(4t - 1) \times (4t - 1)$ identity matrix, and J is the $(4t - 1) \times (4t - 1)$ matrix of all ones. Since also $JA = kJ$, then we know A is the incidence matrix for a $(4t - 1, 4t - 1, 2t - 1, 2t - 1, t - 1)$ block design. ∎

Example 2.3 Consider the normalized Hadamard matrix

$$H_8 = \begin{bmatrix} H_4 & H_4 \\ H_4 & -H_4 \end{bmatrix}$$

of order 8, where H_4 is the normalized Hadamard matrix of order four constructed previously. Using $H = H_8$, Theorem 2.5 states that

$$A = \begin{bmatrix} 0 & 1 & 0 & 1 & 0 & 1 & 0 \\ 1 & 0 & 0 & 1 & 1 & 0 & 0 \\ 0 & 0 & 1 & 1 & 0 & 0 & 1 \\ 1 & 1 & 1 & 0 & 0 & 0 & 0 \\ 0 & 1 & 0 & 0 & 1 & 0 & 1 \\ 1 & 0 & 0 & 0 & 0 & 1 & 1 \\ 0 & 0 & 1 & 0 & 1 & 1 & 0 \end{bmatrix}$$

is the incidence matrix for a $(7, 7, 3, 3, 1)$ block design. Note that this incidence matrix is not the same as the incidence matrix in Example 2.2 for the $(7, 7, 3, 3, 1)$ block design in Example 2.1.   ∎

2.3   Hadamard Matrices with Maple

In this section, we show how Maple can be used to construct the Hadamard matrices H_{2^n} and corresponding block designs discussed in Section 2.2. We consider the design resulting from the incidence matrix in Example 2.3.

Because some of the commands we will use are in the Maple **linalg** linear algebra package, we begin by including this package.

> with(linalg):

Next, we define the Hadamard matrix $H_1 = \begin{bmatrix} 1 \end{bmatrix}$.

> H1 := matrix(1, 1, [1]);

$$H1 := \begin{bmatrix} 1 \end{bmatrix}$$

Recall that the Hadamard matrix H_{2^k} can be constructed as a block matrix from $H_{2^{k-1}}$. Hence, the Hadamard matrices H_2, H_4, and H_8 can be constructed using the Maple **blockmatrix** command as follows.[1]

> H2 := blockmatrix(2, 2, [H1, H1, H1, -H1]);

$$H2 := \begin{bmatrix} 1 & 1 \\ 1 & -1 \end{bmatrix}$$

---

[1] Maple V Release 5 is the first release of Maple that requires brackets [ ] to be included in the **blockmatrix** command around the matrices that form the blocks. For example, to construct the matrix H_2 with an earlier release of Maple, the **blockmatrix** command must be entered as follows.

> H2 := blockmatrix(2, 2, H1, H1, H1, -H1);

```
> H4 := blockmatrix(2, 2, [H2, H2, H2, -H2]);
```

$$H4 := \begin{bmatrix} 1 & 1 & 1 & 1 \\ 1 & -1 & 1 & -1 \\ 1 & 1 & -1 & -1 \\ 1 & -1 & -1 & 1 \end{bmatrix}$$

```
> H8 := blockmatrix(2, 2, [H4, H4, H4, -H4]);
```

$$H8 := \begin{bmatrix} 1 & 1 & 1 & 1 & 1 & 1 & 1 & 1 \\ 1 & -1 & 1 & -1 & 1 & -1 & 1 & -1 \\ 1 & 1 & -1 & -1 & 1 & 1 & -1 & -1 \\ 1 & -1 & -1 & 1 & 1 & -1 & -1 & 1 \\ 1 & 1 & 1 & 1 & -1 & -1 & -1 & -1 \\ 1 & -1 & 1 & -1 & -1 & 1 & -1 & 1 \\ 1 & 1 & -1 & -1 & -1 & -1 & 1 & 1 \\ 1 & -1 & -1 & 1 & -1 & 1 & 1 & -1 \end{bmatrix}$$

The first two parameters in the preceding **blockmatrix** commands are the dimensions of the result in terms of blocks. The remaining parameters are an ordered list of the blocks by rows. Normalized Hadamard matrices of higher orders can be constructed similarly.

We will now construct the incidence matrix shown in Example 2.3 that results from the Hadamard matrix H_8. We first delete the first row and column from H_8 by applying the Maple **delrows** and **delcols** commands as follows.

```
> A := delrows(H8, 1..1):
> A := delcols(A, 1..1):
```

We can then obtain the incidence matrix by changing all negative ones in A to zeros. To do this, we define the following function **f**.

```
> f := x -> if x = -1 then 0 else 1 fi:
```

We then apply the function **f** to each of the entries in A by entering the following **map** command.

```
> A := map(f, A);
```

$$A := \begin{bmatrix} 0 & 1 & 0 & 1 & 0 & 1 & 0 \\ 1 & 0 & 0 & 1 & 1 & 0 & 0 \\ 0 & 0 & 1 & 1 & 0 & 0 & 1 \\ 1 & 1 & 1 & 0 & 0 & 0 & 0 \\ 0 & 1 & 0 & 0 & 1 & 0 & 1 \\ 1 & 0 & 0 & 0 & 0 & 1 & 1 \\ 0 & 0 & 1 & 0 & 1 & 1 & 0 \end{bmatrix}$$

Note that the preceding matrix is the incidence matrix from Example 2.3.

Finally, we will use Maple to list the objects that are contained in each of the blocks in the design. To do this, we first assign the general block design parameters as follows.

```
> v := 7:
> b := 7:
> k := 3:
```

Since each of the blocks in the design will contain k objects, we create the following vector **block** of length k in which to store the objects contained in each block.

```
> block := vector(k);
```

$$block := \text{array}(\, 1..3, [\,]\,)$$

By entering the following commands, we can then see the objects that are contained in each block. In these commands, the outer loop spans the columns of **A**, and the inner loop spans the rows of **A**.

```
> for j from 1 to b do
>      bct := 0:
>      for i from 1 to v do
>          if A[i,j] = 1 then
>              bct := bct + 1;
>              block[bct] := i;
>          fi;
>      od:
>      print('Block    ', j, '    contains objects    ', block);
> od:
```

$$\begin{array}{llll} Block & ,1, & contains\ objects & ,[\,2,4,6\,] \\[4pt] Block & ,2, & contains\ objects & ,[\,1,4,5\,] \\[4pt] Block & ,3, & contains\ objects & ,[\,3,4,7\,] \\[4pt] Block & ,4, & contains\ objects & ,[\,1,2,3\,] \\[4pt] Block & ,5, & contains\ objects & ,[\,2,5,7\,] \\[4pt] Block & ,6, & contains\ objects & ,[\,1,6,7\,] \\[4pt] Block & ,7, & contains\ objects & ,[\,3,5,6\,] \end{array}$$

In the preceding commands, note that we use colons after both **od** statements. This causes Maple to suppress the output (except the output resulting from the **print** command) after each passage through the loop. Note also that, as in Section 1.5, we use back ticks " ' " in the **print** statement.

2.4   Difference Sets

In this section we discuss some techniques for constructing block designs using difference sets. As we will show, these techniques yield block designs with more of a variety of parameters than the designs resulting from Hadamard matrices.

Let G be an abelian group of order v, and let D be a subset of G of order k. If every nonzero element in G can be expressed as the difference of two elements in D in exactly λ ways with $\lambda < k$, then D is called a *difference set* in G, and is described by the parameters (v, k, λ).

Example 2.4 The set $D = \{0, 1, 2, 4, 5, 8, 10\}$ is a $(15, 7, 3)$ difference set in Z_{15}. ∎

Example 2.5 The set $D = \{1, 2, 4\}$ is a $(7, 3, 1)$ difference set in Z_7. If we add each element in Z_7 to each of the elements in D (*i.e.*, if we form the sets $i + D$ for $i = 0, 1, \ldots, 6$), it can easily be verified that the seven resulting sets are the blocks in the block design in Example 2.1 (with $0 \in Z_7$ represented by 7). Hence, the $(7, 3, 1)$ difference set $D = \{1, 2, 4\}$ in Z_7 can be used to construct the $(7, 7, 3, 3, 1)$ block design in Example 2.1. ∎

The fact that a block design results from adding each element in Z_7 to each of the elements in the difference set D in Example 2.5 is guaranteed in general by the following theorem.

Theorem 2.6 *Let* $D = \{d_1, \ldots, d_k\}$ *be a* (v, k, λ) *difference set in the abelian group* $G = \{g_1, \ldots, g_v\}$. *Then the sets*

$$g_i + D = \{g_i + d_1, \ldots, g_i + d_k\}, \qquad i = 1, \ldots, v$$

are the blocks in a (v, v, k, k, λ) *block design.*

Proof. Clearly, there are v objects in the design. Also, the v blocks $g_i + D$ are distinct, for if $g_i + D = g_j + D$ for some $i \neq j$, then $(g_i - g_j) + D = D$. We can then find k differences of elements in D that are equal to $g_i - g_j$, contradicting the assumption that $\lambda < k$. Now, if we add an element in D to each of the elements in G, the result will be the set G. Each element in G will appear k times among the elements $g_i + d_j$ for $i = 1, \ldots, v$ and $j = 1, \ldots, k$. Hence, each element in G will appear in k blocks. Also, by construction, each block will contain k objects. It remains to be shown only that each pair of elements in G appears together in exactly λ blocks. Choose distinct $x, y \in G$. If x, y appear together in some block $g + D$, then

$x = g + d_i$ and $y = g + d_j$ for some i, j. Thus, $x - y = d_i - d_j$, so $x - y$ is the difference of two elements in D. Since D is a (v, k, λ) difference set in G, $x - y$ can be written as the difference of two elements in D in λ ways. And since $x = g + d_i = h + d_i$ implies $g = h$, the difference $d_i - d_j$ cannot come from more than one block. Hence, the pair x, y cannot appear in more than λ blocks. On the other hand, suppose $x - y = d_i - d_j$ for some i, j. Then $x = g + d_i$ for some $g \in G$, and $y = x - (d_i - d_j) = (x - d_i) + d_j = g + d_j$. Thus, x and y appear together in the block $g + D$. Therefore, the pair x, y must appear in at least λ blocks. With our previous result, this implies x and y must appear in exactly λ blocks. ∎

As illustrated in Example 2.5, Theorem 2.6 gives us an easy method for constructing a (v, v, k, k, λ) block design provided we are first able to find a (v, k, λ) difference set. Before discussing how we can construct difference sets, we first generalize them as follows.

Let G be an abelian group of order v, and let $D_1, \ldots, D_t$ be subsets of G of order k. If every nonzero element in G can be expressed as the difference of two elements in the same D_i in exactly λ ways with $\lambda < k$, then the subsets D_i are called *initial blocks* in a *generalized difference set* in G, and are described by the parameters (v, k, λ). Note that for $t = 1$, this definition matches our previous definition of a difference set.

The following theorem generalizes the method given in Theorem 2.6 for constructing block designs from difference sets.

Theorem 2.7 *Let* $D_1, \ldots, D_t$ *be initial blocks in a* (v, k, λ) *generalized difference set in the abelian group* $G = \{g_1, \ldots, g_v\}$. *Then the sets*

$$g_i + D_j, \qquad i = 1, \ldots, v; \quad j = 1, \ldots, t$$

are the blocks in a (v, vt, kt, k, λ) *block design.*

Proof. Exercise. ∎

Example 2.6 The sets $D_1 = \{1, 7, 11\}$, $D_2 = \{2, 14, 3\}$, and $D_3 = \{4, 9, 6\}$ are initial blocks in a $(19, 3, 1)$ generalized difference set in Z_{19}. Theorem 2.7 states that if we add each element in Z_{19} to each of the elements in these initial blocks, the resulting sets will be the blocks in a $(19, 57, 9, 3, 1)$ block design. ∎

As illustrated in Example 2.6, Theorem 2.7 gives us an easy method for constructing a (v, vt, kt, k, λ) block design provided we are first able to find t initial blocks in a (v, k, λ) generalized difference set. The following two propositions give methods for constructing initial blocks in generalized difference sets.

Proposition 2.8 *Suppose $v = 6t + 1 = p^n$ for some prime p and positive integers n and t. Let F be a finite field of order p^n, and choose $a \in F$ such that a is a cyclic generator for F^*. Then the sets*

$$D_i = \left\{a^i, a^{2t+i}, a^{4t+i}\right\}, \qquad i = 0, \ldots, t - 1$$

are initial blocks in a $(6t + 1, 3, 1)$ generalized difference set in F.

Proof. Exercise. (See the proof of Proposition 2.9 below.)   ■

Example 2.7 We can use Proposition 2.8 to construct the initial blocks in Example 2.6 as follows. Let $F = Z_{19}$, and choose cyclic generator $a = 2$ for Z_{19}^*. Since $19 = 6t + 1$ implies $t = 3$, Proposition 2.8 yields three initial blocks. These initial blocks are $D_0 = \left\{2^0, 2^6, 2^{12}\right\} = \{1, 7, 11\}$, $D_1 = \left\{2^1, 2^7, 2^{13}\right\} = \{2, 14, 3\}$, and $D_2 = \left\{2^2, 2^8, 2^{14}\right\} = \{4, 9, 6\}$.   ■

Proposition 2.9 *Suppose $v = 4t + 1 = p^n$ for some prime p and positive integers n and t. Let F be a finite field of order p^n, and choose $a \in F$ such that a is a cyclic generator for F^*. Then the sets*

$$D_i = \left\{a^i, a^{t+i}, a^{2t+i}, a^{3t+i}\right\}, \qquad i = 0, \ldots, t - 1$$

are initial blocks in a $(4t + 1, 4, 3)$ generalized difference set in F.

Proof. Since a is a cyclic generator for F^*, the order of a is $4t$. Hence, $a^{4t} = 1$, and $a^{2t} \neq 1$. Also, $a^{4t} - 1 = (a^{2t} - 1)(a^{2t} + 1) = 0$ implies $a^{2t} = -1$. Furthermore, $a^t - 1 \neq 0$, so $a^t - 1 = a^s$ for some s between 1 and $4t$. Forming all possible differences from the sets $\pm a^i(a^t - 1)$, $\pm a^i(a^{2t} - 1)$, $\pm a^i(a^{2t} - a^t)$, $\pm a^i(a^{3t} - 1)$, $\pm a^i(a^{3t} - a^t)$, and $\pm a^i(a^{3t} - a^{2t})$, we obtain the following.

$$
\begin{array}{lcllcl}
\pm a^i(a^t - 1) & = & \pm a^i(a^s) & = & a^{i+s}, & a^{i+s+2t} \\
\pm a^i(a^{2t} - 1) & = & \pm a^i(2a^{2t}) & = & 2a^{i+2t}, & 2a^i \\
\pm a^i(a^{2t} - a^t) & = & \pm a^i a^t(a^t - 1) & = & a^{i+t+s}, & a^{i+3t+s} \\
\pm a^i(a^{3t} - 1) & = & \pm a^i a^{3t}(1 - a^t) & = & a^{i+t+s}, & a^{i+3t+s} \\
\pm a^i(a^{3t} - a^t) & = & \pm a^i a^t(2a^{2t}) & = & 2a^{i+3t}, & 2a^{i+t} \\
\pm a^i(a^{3t} - a^{2t}) & = & \pm a^i a^{2t}(a^s) & = & a^{i+2t+s}, & a^{i+s}
\end{array}
$$

Multiplication by a^s and 2 are bijections, so these elements can be canceled from the preceding expressions. The only remaining elements are a^i, a^{t+i}, a^{2t+i}, and a^{3t+i} for $i = 0, \ldots, t - 1$ repeated three times each. Since these are all of the elements in F^*, then $\lambda = 3$.   ■

Example 2.8 Suppose we wish to obtain a fair and reasonable comparison of 9 cars by evaluating the opinions of 18 consumers. We can use Proposition 2.9 to construct a block design for this comparison as follows. We first need a finite field F of order 9 to represent the cars. For F, we will use the field of order 9 constructed in Example 1.10. For the cyclic generator a for F^*, we will use the element $x \in F$. Since $9 = 4t + 1$ implies $t = 2$, Proposition 2.9 yields two initial blocks in a $(9, 4, 3)$ generalized difference set in F. These initial blocks are $D_0 = \left\{1, x^2, x^4, x^6\right\} = \{1, 2x+1, 2, x+2\}$ and $D_1 = \left\{x, x^3, x^5, x^7\right\} = \{x, 2x+2, 2x, x+1\}$. Theorem 2.7 states that if we add each element in F to each of the elements in these initial blocks, the resulting sets will be the blocks in a $(9, 18, 8, 4, 3)$ block design. The blocks in this design are listed at the end of Section 2.5. Note that in this block design, each car is tested 8 times, each consumer tests 4 cars, and each pair of cars is tested by the same consumer 3 times. ■

2.5 Difference Sets with Maple

In this section, we show how Maple can be used to construct the initial blocks and corresponding block designs discussed in Section 2.4. We consider the design resulting from the initial blocks in Example 2.8.

We begin by including the Maple **linalg** package and entering the primitive polynomial $f(x) = x^2 + x + 2 \in Z_3[x]$ used to construct the elements in the finite field F.

```
> with(linalg):
> f := x -> x^2 + x + 2:
> Primitive(f(x)) mod 3;
```

$$true$$

Recall that since $v = 4t + 1 = 9$ implies $t = 2$, there will be 2 initial blocks. We define this parameter next.

```
> t := 2:
```

Because the field elements are the objects that will fill the blocks, we must store these elements in a way such that they can be recalled. We will do this by storing the field elements in a vector. We first initialize a vector with the same number of positions as the number of field elements.

```
> field := vector(4*t+1);
```

$$field := \mathrm{array}(\,1..9, [\,]\,)$$

Then by entering the following commands, we generate and store the field elements in the vector **field**. Note the bracket [ ] syntax for accessing the positions in **field**.

```
> for i from 1 to 4*t do
>     field[i] := Powmod(x, i, f(x), x) mod 3:
> od:
> field[4*t+1] := 0:
```

We can view the entries in the vector **field** by entering the following **evalm** command.

```
> evalm(field);
```

$$[x,\, 2x+1,\, 2x+2,\, 2,\, 2x,\, x+2,\, x+1,\, 1,\, 0]$$

Next, we define the number $k = 4$ of objects contained in each of the initial blocks and blocks in the design, and create a vector in which to store the initial blocks.

```
> k := 4:
> initblock := vector(k);
```

$$initblock := \mathrm{array}(\,1..4, [\,]\,)$$

We can then generate and display the initial blocks by entering the following commands. In these commands, the outer loop spans the initial blocks while the inner loop constructs the entries in each one.

```
> for i from 0 to t-1 do
>     for j from 1 to k do
>         initblock[j] := Powmod(x, (j-1)*t+i, f(x), x) mod 3;
>     od:
>     print('Initial Block   ', i, '   is   ', initblock);
> od:
```

$$\textit{Initial Block}\qquad,0,\quad \textit{is}\quad,[1,\, 2x+1,\, 2,\, x+2]$$

$$\textit{Initial Block}\qquad,1,\quad \textit{is}\quad,[x,\, 2x+2,\, 2x,\, x+1]$$

In order to construct all of the blocks in the design, we first create a vector in which to store the blocks, and initialize a counter **bct** we will use to number the blocks.

```
> block := vector(k):
> bct := 0:
```

We can then generate and display all of the blocks in the design by entering the following commands. In these commands, the outer loop spans the initial blocks while the first inner loop constructs the entries in each one.

The last two inner loops add each field element to each of the elements in
the initial blocks, thus yielding the blocks in the design.

```
> for i from 0 to t-1 do
>       for j from 1 to k do
>           initblock[j] := Powmod(x, (j-1)*t+i, f(x), x) mod 3;
>       od:
>       for j from 1 to 4*t+1 do
>           for h from 1 to k do
>               block[h] := (field[j] + initblock[h]) mod 3;
>           od:
>           bct := bct + 1;
>           print('Block    ', bct, '    is    ', block);
>       od:
> od:
```

$$Block \quad , 1, \quad is \quad , [x+1, 1, x+2, 2x+2]$$

$$Block \quad , 2, \quad is \quad , [2x+2, x+2, 2x, 0]$$

$$Block \quad , 3, \quad is \quad , [2x, x, 2x+1, 1]$$

$$Block \quad , 4, \quad is \quad , [0, 2x, 1, x+1]$$

$$Block \quad , 5, \quad is \quad , [2x+1, x+1, 2x+2, 2]$$

$$Block \quad , 6, \quad is \quad , [x, 0, x+1, 2x+1]$$

$$Block \quad , 7, \quad is \quad , [x+2, 2, x, 2x]$$

$$Block \quad , 8, \quad is \quad , [2, 2x+2, 0, x]$$

$$Block \quad , 9, \quad is \quad , [1, 2x+1, 2, x+2]$$

$$Block \quad , 10, \quad is \quad , [2x, 2, 0, 2x+1]$$

$$Block \quad , 11, \quad is \quad , [1, x, x+1, 2]$$

$$Block \quad , 12, \quad is \quad , [2, x+1, x+2, 0]$$

$$Block \quad , 13, \quad is \quad , [x+2, 2x+1, 2x+2, x]$$

$$Block \quad , 14, \quad is \quad , [0, x+2, x, 1]$$

$$Block \quad , 15, \quad is \quad , [2x+2, 1, 2, 2x]$$

$$Block \quad , 16, \quad is \quad , [2x+1, 0, 1, 2x+2]$$

$$Block \quad , 17, \quad is \quad , [x+1, 2x, 2x+1, x+2]$$

$$Block \quad , 18, \quad is \quad , [x, 2x+2, 2x, x+1]$$

Written Exercises

1. Suppose a magazine editor wishes to obtain a comparison of 15 cars by evaluating the opinions of 15 consumers. Construct a block design for this comparison. List the block design parameters.

2. Construct two different block designs with $v = 13$ objects. List the block design parameters for each one.

3. Suppose a magazine editor wishes to obtain a comparison of 25 cars by evaluating the opinions of a certain number of consumers after each of the consumers tests 3 of the cars. Construct a block design for this comparison. List the block design parameters, and state what each parameter represents.

4. Repeat Written Exercise 3 if the editor decides to have each of the consumers test 4 of the cars instead of 3.

5. Repeat Written Exercise 3 if the editor decides to compare only 7 cars instead of 25. (Assume each consumer still tests 3 of the cars.)

6. Show that if H is a Hadamard matrix, then so is $\begin{bmatrix} H & H \\ H & -H \end{bmatrix}$.

7. Prove Theorem 2.7.

8. Prove Proposition 2.8.

Maple Exercises

1. Suppose a magazine editor wishes to obtain a comparison of 31 cars by evaluating the opinions of 31 consumers. Construct a block design for this comparison. List the block design parameters.

2. Suppose a magazine editor wishes to obtain a comparison of 81 types of candy by evaluating the opinions of a certain number of children after each child samples 4 of the types of candy. Construct the initial blocks in a block design for this comparison. List the block design parameters, and state what each parameter represents.

3. Construct two different block designs with $v = 121$ objects. (Construct initial blocks only if you use Propositions 2.8 or 2.9.) List the block design parameters for each one.

4. Construct two different block designs with $v = 127$ objects. (Construct initial blocks only if you use Propositions 2.8 or 2.9.) List the block design parameters for each one.

Chapter 3

Error-Correcting Codes

In the next three chapters we discuss several types of error-correcting codes. A *code* is a set of messages called *codewords* that can be transmitted between two parties. An "error-correcting" code is a code for which it is sometimes possible to detect and correct errors that occur during transmission of the codewords. Some applications of error-correcting codes include correction of errors that occur in information transmitted via the Internet, data stored in a computer, and music encoded on a compact disc. Error-correcting codes can also be used to correct errors that occur in information transmitted through space. For example, we mention in Section 3.3 how an error-correcting code was used in the Mariner 9 space probe when it returned photographs of Mars to Earth in 1972.

3.1   General Properties of Codes

In this chapter we consider some types of codes in which the codewords are vectors of a fixed length over Z_2. We will denote the space of vectors of length n over Z_2 as Z_2^n. Hence, the codes we consider in this chapter will be subsets of Z_2^n for some n. A code C in Z_2^n is not required to be a subspace of Z_2^n. If C is a subspace of Z_2^n, then we call C a *linear* code. We discuss linear codes beginning in Section 3.5 and continuing in Chapters 4 and 5.

The way we will tell in general if an error occurred during the transmission of a codeword in a code C is by determining if the received vector is in C. Thus, because our goal is to be able to detect and correct errors

43

in received vectors, not all vectors in Z_2^n for some n can be codewords in a code C. In general, we will use the "nearest neighbor policy" to correct a received vector that contains errors. This means that we will assume the fewest possible number of errors, and correct the received vector to the codeword from which it differs in the fewest positions. This method of error correction is limited, for there is not always a unique codeword that differs from a received vector in the fewest positions.

Example 3.1 Consider the code $C = \{(1010), (1110), (0011)\}$ in Z_2^4. Suppose a codeword in C is transmitted and we receive the vector $r_1 = (0110)$. A quick search of C reveals that $c = (1110)$ is the codeword from which r_1 differs in the fewest positions. Hence, we would correct r_1 to c, and assume that the error in r_1 is $e = r_1 - c = (1000)$. Now, suppose a codeword in C is transmitted and we receive the vector $r_2 = (0010)$. Since two of the codewords in C differ from r_2 in only one position, we cannot uniquely correct r_2 using the nearest neighbor policy. Therefore, in this code C, we are not guaranteed to be able to uniquely correct a received vector in Z_2^4 even if the received vector contains only a single error. ∎

To make the nearest neighbor policy error correction method more precise, we make the following definition. Let C be a code in Z_2^n. For vectors $x, y \in C$, we define the *Hamming distance* $d(x, y)$ from x to y to be the number of positions in which x and y differ. Hence, if $x = (x_1, \ldots, x_n)$ and $y = (y_1, \ldots, y_n)$, then $d(x, y) = \sum_{i=1}^{n} |x_i - y_i|$. We will call the smallest Hamming distance between any two codewords in a code C the *minimum distance* of C. We will denote this minimum distance by $d(C)$, or just d if there is no confusion regarding the code to which we are referring. For example, for the code C in Example 3.1, $d = 1$.

Determining the number of errors that are guaranteed to be uniquely correctable in a given code is an important part of coding theory. To do this in general, consider the following. For $x \in Z_2^n$ and positive integer r, let $S_r(x) = \{y \in Z_2^n \mid d(x, y) \leq r\}$. In standard terminology, $S_r(x)$ is called the *ball of radius r around x*. Let C be a code with minimum distance d, and let t be the largest integer such that $t < \frac{d}{2}$. Then $S_t(x) \cap S_t(y)$ is empty for every pair x, y of distinct codewords in C. If z is a received vector in Z_2^n with $d(u, z) \leq t$ for some $u \in C$, then $z \in S_t(u)$ and $z \notin S_t(v)$ for all other $v \in C$. That is, if a received vector $z \in Z_2^n$ differs from a codeword $u \in C$ in t or fewer positions, then every other codeword in C will differ from z in more than t positions. Thus, the nearest neighbor policy will always allow t or fewer errors to be corrected in the code. The code C is said to be *t-error correcting*.

Example 3.2 Let $C = \{(00000000), (11100011), (00011111), (11111100)\}$. It can easily be seen that the minimum distance of C is $d = 5$. Since $t = 2$ is the largest integer such that $t < \frac{d}{2}$, then C is 2-error correcting. ∎

Suppose C is a t-error correcting code in $V = Z_2^n$. We now address the problem of determining the number of vectors in V that are guaranteed to be correctable in C. Note first that for any $x \in V$, there are $\binom{n}{i}$ vectors in V that differ from x in exactly i positions. Also, any vector in V that differs from x in i positions will be in $S_t(x)$ provided $i \leq t$. Hence, the number of vectors in $S_t(x)$ will be $\binom{n}{0} + \binom{n}{1} + \cdots + \binom{n}{t}$. To determine the number of vectors in V that are guaranteed to be correctable in C, we must only count the number of vectors in $S_t(x)$ as x ranges through the codewords in C. Since the sets $S_t(x)$ are pairwise disjoint, the number of vectors in V that differ from one of the codewords in C in t or fewer positions, and are consequently guaranteed to be uniquely correctable in C, is $|C| \cdot \left[\binom{n}{0} + \binom{n}{1} + \cdots + \binom{n}{t} \right]$. The fact that $|V| = 2^n$ then yields the following theorem, which gives a bound on the number of vectors in Z_2^n that are guaranteed to be correctable in a t-error correcting code in Z_2^n. This bound is called the *Hamming bound*.

Theorem 3.1 *Suppose C is a t-error correcting code in Z_2^n. Then*

$$|C| \cdot \left[\binom{n}{0} + \binom{n}{1} + \cdots + \binom{n}{t} \right] \leq 2^n.$$

A code C in Z_2^n is said to be *perfect* if every vector in Z_2^n is guaranteed to be correctable in C. That is, a code C in Z_2^n is perfect if the inequality in Theorem 3.1 with C is an equality. For the code C in Example 3.2, the factors in this inequality are $|C| = 4$, $\binom{8}{0} + \binom{8}{1} + \binom{8}{2} = 37$, and $2^8 = 256$. Thus, for the code C in Example 3.2, 108 of the vectors in Z_2^8 are not guaranteed to be uniquely correctable in C (some may, however, still be "closest" to a unique codeword). Therefore, this code is far from perfect. In Sections 3.5 and 3.6 we discuss a class of codes called *Hamming codes* that are perfect.

In practice, it is often desirable to construct codes that have a large number of codewords and are guaranteed to correct a large number of errors. However, the number of errors guaranteed to be correctable in a code is obviously related to the number of codewords in the code. Indeed, to construct a t-error correcting code C in Z_2^n for fixed values of n and t such that $|C|$ is maximized has been an important problem of recent mathematical interest. An equivalent problem is to find the maximum number of points in Z_2^n such that the balls of a fixed radius around the points can be arranged in the space without intersecting. This type of problem is called a *sphere packing* problem.

For the remainder of this chapter and the two subsequent chapters we will discuss several methods for constructing various types of codes and correcting errors in these codes. To facilitate this, we will establish a set of parameters to use in describing codes. We will describe a code by the parameters (n, d) if the codewords in the code are of length n positions and the code has minimum distance d.

3.2 Hadamard Codes

In Section 2.2 we showed that for certain values of v, k, and λ, it is possible to use a Hadamard matrix to construct an incidence matrix for a (v, v, k, k, λ) block design. The following theorem states that the rows of such an incidence matrix form the codewords in an error-correcting code.

Theorem 3.2 *Suppose A is an incidence matrix for a (v, v, k, k, λ) block design. Then the rows of A form a $(v, 2(k - \lambda))$ code with v codewords.*

Proof. There are v positions in each of the v rows of A. Hence, the rows of A form a code with v codewords each of length v positions. It remains to be shown only that the minimum distance of this code is $2(k - \lambda)$. Consider rows R_1 and R_2 in A. Since each row of A contains ones in k positions, and each pair of rows of A contains ones in λ positions in common, there will be $k - \lambda$ positions in which R_1 contains a one and R_2 contains a zero, and $k - \lambda$ positions in which these elements are reversed. This yields $2(k - \lambda)$ positions in which R_1 and R_2 differ. ∎

Example 3.3 Theorem 3.2 states that the rows of the incidence matrix A in Example 2.3 form a $(7, 4)$ code with 7 codewords. ∎

In Theorem 2.5, we showed that a normalized Hadamard matrix H of order $4m \geq 8$ can be used to construct an incidence matrix for a $(4m-1, 4m-1, 2m-1, 2m-1, m-1)$ block design. Theorem 3.2 states that the rows of such an incidence matrix A form codewords of length $4m - 1$

positions in a code with minimum distance $d = 2((2m-1)-(m-1)) = 2m$ and $4m - 1$ codewords. Recall that each of the rows of A will contain $2m$ zeros and $2m - 1$ ones. Hence, there will be $2m$ positions in which the vector $(1\ 1\ \cdots\ 1)$ of length $4m-1$ positions differs from each of the rows of A. Thus, by including the vector $(1\ 1\ \cdots\ 1)$ of length $4m-1$ positions with the rows of A, we obtain a $(4m-1, 2m)$ code with $4m$ codewords. And no more vectors can be included in this code without decreasing the minimum distance of the code (see Corollary 3.4). Because these $(4m-1, 2m)$ codes with $4m$ codewords are constructed from Hadamard matrices, we will call them *Hadamard* codes.

We close this section by proving the following theorem and corollary which verify the fact mentioned above that no vectors can be joined to the codewords in a Hadamard code without decreasing the minimum distance of the code.

Theorem 3.3 *Let r be the number of codewords in a code with parameters (n, d) for some n, d with $d > \frac{n}{2}$. Then $r \leq \frac{2d}{2d-n}$.*

Proof. Let $A = (a_{ij})$ be an $r \times n$ matrix with the codewords as rows, and let $S = \sum_{u,v} d(u, v)$ for all distinct pairs u, v of codewords. Now, $d(u, v) \geq d$ for all pairs u, v of codewords. Hence, $S \geq \binom{r}{2} d = \frac{r(r-1)}{2} d$. Let $t_0^{(i)}$ and $t_1^{(i)}$ be the number of times that 0 and 1 appear in the ith column of A, respectively. Then $t_1^{(i)} + t_0^{(i)} = r$ for all i. Also,

$$S = \sum_{\Omega} \sum_{j} |a_{ij} - a_{kj}| = \sum_{j} \sum_{\Omega} |a_{ij} - a_{kj}|,$$

where Ω is the set of all distinct pairs of rows of A. For each j, $\sum_{\Omega} |a_{ij} - a_{kj}|$ is equal to the number of times that any two rows of A contain differing entries in the jth position. This number is $t_0^{(j)} t_1^{(j)}$, so $S = \sum_{j} t_0^{(j)} \left(r - t_0^{(j)} \right)$. To find an upper bound on $t_0^{(j)} t_1^{(j)}$, we consider the function $f(x) = x(r - x)$ for $0 \leq x \leq r$. Note that $f(x)$ is maximized at the point $(x, f(x)) = \left(\frac{r}{2}, \frac{r^2}{4} \right)$. Hence, $t_0^{(j)} t_1^{(j)} \leq \frac{r^2}{4}$, and $S \leq \frac{nr^2}{4}$. Thus, $\frac{r(r-1)d}{2} \leq \frac{nr^2}{4}$, and $r \left(d - \frac{n}{2} \right) \leq d$. Therefore, $r \leq \frac{d}{d - \frac{n}{2}} = \frac{2d}{2d-n}$. ∎

Corollary 3.4 *Let r be the number of codewords in a code with parameters $(4m-1, 2m)$ for some m. Then $r \leq 4m$.*

Proof. Exercise. ∎

3.3   Reed-Muller Codes

In Section 3.2, we showed that normalized Hadamard matrices can be used to construct error-correcting codes we called Hadamard codes. We also showed that the number of codewords in a Hadamard code is maximal in the sense that no vectors can be included in the code without decreasing the minimum distance of the code. As a consequence of the following theorem, by increasing the length of the codewords in a Hadamard code by one position, we can double the number of codewords in the code without decreasing the minimum distance of the code.

Theorem 3.5 *Suppose A is the incidence matrix constructed from a normalized Hadamard matrix of order $4m$, and let B be the matrix that results from interchanging all zeros and ones in A. Let $\mathcal{A}$ be the matrix obtained by placing a one in front of all of the rows of A, and let $\mathcal{B}$ be the matrix obtained by placing a zero in front of all of the rows of B. Then the rows of $\mathcal{A}$ and $\mathcal{B}$ taken together form a $(4m, 2m)$ code with $8m - 2$ codewords.*

Proof. Exercise.   ∎

Each of the rows in the matrices $\mathcal{A}$ and $\mathcal{B}$ in Theorem 3.5 will contain $2m$ zeros and $2m$ ones. Hence, there will be $2m$ positions in which both of the vectors $(0\ 0\ \cdots\ 0)$ and $(1\ 1\ \cdots\ 1)$ of length $4m$ positions differ from each of the rows of $\mathcal{A}$ and $\mathcal{B}$. Thus, by including the vectors $(0\ 0\ \cdots\ 0)$ and $(1\ 1\ \cdots\ 1)$ of length $4m$ positions with the rows of $\mathcal{A}$ and $\mathcal{B}$, we obtain a $(4m, 2m)$ code with $8m$ codewords. And no more vectors can be included in this code without decreasing the minimum distance of the code. These $(4m, 2m)$ codes with $8m$ codewords are called *Reed-Muller* codes.

A Reed-Muller code was used in the Mariner 9 space probe when it returned photographs of Mars to Earth in 1972. The specific code used in the space probe was the $(32, 16)$ Reed-Muller code with 64 codewords constructed using the normalized Hadamard matrix H_{32} of order 32 (see Maple Exercise 1). Before being transmitted, each photograph was broken down into a collection of very small dots. Each dot was then assigned one of 64 levels of grayness and encoded into one of the 64 codewords.

3.4   Reed-Muller Codes with Maple

In this section, we show how Maple can be used to construct and correct errors in the $(16, 8)$ Reed-Muller code.

We begin by generating the normalized Hadamard matrix H_{16} of order 16 used to construct the code.[1]

```
> with(linalg):
> H1 := matrix(1, 1, [1]):
> H2 := blockmatrix(2, 2, [H1, H1, H1, -H1]):
> H4 := blockmatrix(2, 2, [H2, H2, H2, -H2]):
> H8 := blockmatrix(2, 2, [H4, H4, H4, -H4]):
> H16 := blockmatrix(2, 2, [H8, H8, H8, -H8]):
```

We can then construct the incidence matrix A that results from H_{16} as follows.

```
> A := delrows(H16, 1..1):
> A := delcols(A, 1..1):
> f := x -> if x = -1 then 0 else 1 fi:
> A := map(f, A);
```

$$A := \begin{bmatrix}
0 & 1 & 0 & 1 & 0 & 1 & 0 & 1 & 0 & 1 & 0 & 1 & 0 & 1 & 0 \\
1 & 0 & 0 & 1 & 1 & 0 & 0 & 1 & 1 & 0 & 0 & 1 & 1 & 0 & 0 \\
0 & 0 & 1 & 1 & 0 & 0 & 1 & 1 & 0 & 0 & 1 & 1 & 0 & 0 & 1 \\
1 & 1 & 1 & 0 & 0 & 0 & 0 & 1 & 1 & 1 & 1 & 0 & 0 & 0 & 0 \\
0 & 1 & 0 & 0 & 1 & 0 & 1 & 1 & 0 & 1 & 0 & 0 & 1 & 0 & 1 \\
1 & 0 & 0 & 0 & 0 & 1 & 1 & 1 & 1 & 0 & 0 & 0 & 0 & 1 & 1 \\
0 & 0 & 1 & 0 & 1 & 1 & 0 & 1 & 0 & 0 & 1 & 0 & 1 & 1 & 0 \\
1 & 1 & 1 & 1 & 1 & 1 & 1 & 0 & 0 & 0 & 0 & 0 & 0 & 0 & 0 \\
0 & 1 & 0 & 1 & 0 & 1 & 0 & 0 & 1 & 0 & 1 & 0 & 1 & 0 & 1 \\
1 & 0 & 0 & 1 & 1 & 0 & 0 & 0 & 0 & 1 & 1 & 0 & 0 & 1 & 1 \\
0 & 0 & 1 & 1 & 0 & 0 & 1 & 0 & 1 & 1 & 0 & 0 & 1 & 1 & 0 \\
1 & 1 & 1 & 0 & 0 & 0 & 0 & 0 & 0 & 0 & 0 & 1 & 1 & 1 & 1 \\
0 & 1 & 0 & 0 & 1 & 0 & 1 & 0 & 1 & 0 & 1 & 1 & 0 & 1 & 0 \\
1 & 0 & 0 & 0 & 0 & 1 & 1 & 0 & 0 & 1 & 1 & 1 & 1 & 0 & 0 \\
0 & 0 & 1 & 0 & 1 & 1 & 0 & 0 & 1 & 1 & 0 & 1 & 0 & 0 & 1
\end{bmatrix}$$

Next, we construct the matrix B from Theorem 3.5 by interchanging all zeros and ones in the matrix A. To do this, we define and apply the following function g to the entries in A.

```
> g := x -> if x = 0 then 1 else 0 fi:
> B := map(g, A);
```

---

[1] See footnote p. 33 regarding the Maple **blockmatrix** command.

$$B := \begin{bmatrix}
1 & 0 & 1 & 0 & 1 & 0 & 1 & 0 & 1 & 0 & 1 & 0 & 1 & 0 & 1 \\
0 & 1 & 1 & 0 & 0 & 1 & 1 & 0 & 0 & 1 & 1 & 0 & 0 & 1 & 1 \\
1 & 1 & 0 & 0 & 1 & 1 & 0 & 0 & 1 & 1 & 0 & 0 & 1 & 1 & 0 \\
0 & 0 & 0 & 1 & 1 & 1 & 1 & 0 & 0 & 0 & 0 & 1 & 1 & 1 & 1 \\
1 & 0 & 1 & 1 & 0 & 1 & 0 & 0 & 1 & 0 & 1 & 1 & 0 & 1 & 0 \\
0 & 1 & 1 & 1 & 1 & 0 & 0 & 0 & 0 & 1 & 1 & 1 & 1 & 0 & 0 \\
1 & 1 & 0 & 1 & 0 & 0 & 1 & 0 & 1 & 1 & 0 & 1 & 0 & 0 & 1 \\
0 & 0 & 0 & 0 & 0 & 0 & 0 & 1 & 1 & 1 & 1 & 1 & 1 & 1 & 1 \\
1 & 0 & 1 & 0 & 1 & 0 & 1 & 1 & 0 & 1 & 0 & 1 & 0 & 1 & 0 \\
0 & 1 & 1 & 0 & 0 & 1 & 1 & 1 & 1 & 0 & 0 & 1 & 1 & 0 & 0 \\
1 & 1 & 0 & 0 & 1 & 1 & 0 & 1 & 0 & 0 & 1 & 1 & 0 & 0 & 1 \\
0 & 0 & 0 & 1 & 1 & 1 & 1 & 1 & 1 & 1 & 1 & 0 & 0 & 0 & 0 \\
1 & 0 & 1 & 1 & 0 & 1 & 0 & 1 & 0 & 1 & 0 & 0 & 1 & 0 & 1 \\
0 & 1 & 1 & 1 & 1 & 0 & 0 & 1 & 1 & 0 & 0 & 0 & 0 & 1 & 1 \\
1 & 1 & 0 & 1 & 0 & 0 & 1 & 1 & 0 & 0 & 1 & 0 & 1 & 1 & 0
\end{bmatrix}$$

We now construct the matrices $\mathcal{A}$ and $\mathcal{B}$ from Theorem 3.5. Recall that to construct $\mathcal{A}$ and $\mathcal{B}$, we must place a one in front of all of the rows of A and a zero in front of all of the rows of B. To do this, we first define the following vectors colA and colB.

```
> colA := vector(rowdim(A), 1);
```

$$colA := [1, 1, 1, 1, 1, 1, 1, 1, 1, 1, 1, 1, 1, 1, 1]$$

```
> colB := vector(rowdim(B), 0);
```

$$colB := [0, 0, 0, 0, 0, 0, 0, 0, 0, 0, 0, 0, 0, 0, 0]$$

By construction, the preceding vectors both have the same number of positions as the number of rows in A or B. Hence, by placing the vectors colA and colB as columns in front of the matrices A and B respectively, we will obtain the matrices $\mathcal{A}$ and $\mathcal{B}$. We can do this using the Maple **augment** command as follows.

```
> scriptA := augment(colA, A):
> scriptB := augment(colB, B):
```

The rows of the matrices scriptA and scriptB taken together form all but two of the codewords in the $(16, 8)$ Reed-Muller code. The two codewords not included in the rows of these matrices are the vectors $(0\ 0\ \cdots\ 0)$ and $(1\ 1\ \cdots\ 1)$ of length 16 positions. We create these vectors next.

```
> v_zero := vector(coldim(scriptB), 0):
> v_one := vector(coldim(scriptB), 1):
```

We can then view the codewords in the $(16, 8)$ Reed-Muller code by using the Maple **stackmatrix** command[2] as follows to stack the matrices $\mathcal{A}$ and $\mathcal{B}$ and the vectors **v_zero** and **v_one**.

```
> cw := stackmatrix(scriptA, scriptB, v_zero, v_one);
```

$$cw := \begin{bmatrix}
1 & 0 & 1 & 0 & 1 & 0 & 1 & 0 & 1 & 0 & 1 & 0 & 1 & 0 & 1 & 0 \\
1 & 1 & 0 & 0 & 1 & 1 & 0 & 0 & 1 & 1 & 0 & 0 & 1 & 1 & 0 & 0 \\
1 & 0 & 0 & 1 & 1 & 0 & 0 & 1 & 1 & 0 & 0 & 1 & 1 & 0 & 0 & 1 \\
1 & 1 & 1 & 1 & 0 & 0 & 0 & 0 & 1 & 1 & 1 & 1 & 0 & 0 & 0 & 0 \\
1 & 0 & 1 & 0 & 0 & 1 & 0 & 1 & 1 & 0 & 1 & 0 & 0 & 1 & 0 & 1 \\
1 & 1 & 0 & 0 & 0 & 0 & 1 & 1 & 1 & 1 & 0 & 0 & 0 & 0 & 1 & 1 \\
1 & 0 & 0 & 1 & 0 & 1 & 1 & 0 & 1 & 0 & 0 & 1 & 0 & 1 & 1 & 0 \\
1 & 1 & 1 & 1 & 1 & 1 & 1 & 1 & 0 & 0 & 0 & 0 & 0 & 0 & 0 & 0 \\
1 & 0 & 1 & 0 & 1 & 0 & 1 & 0 & 0 & 1 & 0 & 1 & 0 & 1 & 0 & 1 \\
1 & 1 & 0 & 0 & 1 & 1 & 0 & 0 & 0 & 0 & 1 & 1 & 0 & 0 & 1 & 1 \\
1 & 0 & 0 & 1 & 1 & 0 & 0 & 1 & 0 & 1 & 1 & 0 & 0 & 1 & 1 & 0 \\
1 & 1 & 1 & 1 & 0 & 0 & 0 & 0 & 0 & 0 & 0 & 0 & 1 & 1 & 1 & 1 \\
1 & 0 & 1 & 0 & 0 & 1 & 0 & 1 & 0 & 1 & 0 & 1 & 1 & 0 & 1 & 0 \\
1 & 1 & 0 & 0 & 0 & 0 & 1 & 1 & 0 & 0 & 1 & 1 & 1 & 1 & 0 & 0 \\
1 & 0 & 0 & 1 & 0 & 1 & 1 & 0 & 0 & 1 & 1 & 0 & 1 & 0 & 0 & 1 \\
0 & 1 & 0 & 1 & 0 & 1 & 0 & 1 & 0 & 1 & 0 & 1 & 0 & 1 & 0 & 1 \\
0 & 0 & 1 & 1 & 0 & 0 & 1 & 1 & 0 & 0 & 1 & 1 & 0 & 0 & 1 & 1 \\
0 & 1 & 1 & 0 & 0 & 1 & 1 & 0 & 0 & 1 & 1 & 0 & 0 & 1 & 1 & 0 \\
0 & 0 & 0 & 0 & 1 & 1 & 1 & 1 & 0 & 0 & 0 & 0 & 1 & 1 & 1 & 1 \\
0 & 1 & 0 & 1 & 1 & 0 & 1 & 0 & 0 & 1 & 0 & 1 & 1 & 0 & 1 & 0 \\
0 & 0 & 1 & 1 & 1 & 1 & 0 & 0 & 0 & 0 & 1 & 1 & 1 & 1 & 0 & 0 \\
0 & 1 & 1 & 0 & 1 & 0 & 0 & 1 & 0 & 1 & 1 & 0 & 1 & 0 & 0 & 1 \\
0 & 0 & 0 & 0 & 0 & 0 & 0 & 0 & 1 & 1 & 1 & 1 & 1 & 1 & 1 & 1 \\
0 & 1 & 0 & 1 & 0 & 1 & 0 & 1 & 1 & 0 & 1 & 0 & 1 & 0 & 1 & 0 \\
0 & 0 & 1 & 1 & 0 & 0 & 1 & 1 & 1 & 1 & 0 & 0 & 1 & 1 & 0 & 0 \\
0 & 1 & 1 & 0 & 0 & 1 & 1 & 0 & 1 & 0 & 0 & 1 & 1 & 0 & 0 & 1 \\
0 & 0 & 0 & 0 & 1 & 1 & 1 & 1 & 1 & 1 & 1 & 1 & 0 & 0 & 0 & 0 \\
0 & 1 & 0 & 1 & 1 & 0 & 1 & 0 & 1 & 0 & 1 & 0 & 0 & 1 & 0 & 1 \\
0 & 0 & 1 & 1 & 1 & 1 & 0 & 0 & 1 & 1 & 0 & 0 & 0 & 0 & 1 & 1 \\
0 & 1 & 1 & 0 & 1 & 0 & 0 & 1 & 1 & 0 & 0 & 1 & 0 & 1 & 1 & 0 \\
0 & 0 & 0 & 0 & 0 & 0 & 0 & 0 & 0 & 0 & 0 & 0 & 0 & 0 & 0 & 0 \\
1 & 1 & 1 & 1 & 1 & 1 & 1 & 1 & 1 & 1 & 1 & 1 & 1 & 1 & 1 & 1
\end{bmatrix}$$

Recall that the $(4m, 2m)$ Reed-Muller code contains $8m$ codewords. Hence,

---

[2]Maple V Release 5 is the first release of Maple that uses **stackmatrix** to stack matrices and vectors vertically. Earlier releases of Maple use the **stack** command to accomplish this. For example, with an earlier release of Maple, we would construct the matrix **cw** by entering the following command.

```
> cw := stack(scriptA, scriptB, v_zero, v_one);
```

there should be 32 rows in the preceding matrix. We can verify this as follows.

> rowdim(cw);

32

We now show how Maple can be used to correct a vector in Z_2^{16} in the $(16, 8)$ Reed-Muller code. Note first that since in general the $(4m, 2m)$ Reed-Muller code is $(m - 1)$-error correcting, the $(16, 8)$ Reed-Muller code will be 3-error correcting. Suppose a codeword in the $(16, 8)$ Reed-Muller code is transmitted and we receive the following vector.

> r := vector([1, 0, 1, 0, 1, 0, 0, 1, 0, 1, 1, 0, 1, 0, 0, 0]):

We can then use the following commands to determine if this received vector contains a correctable error. We first define the general Reed-Muller parameter $m = 4$, and then define two additional parameters fc and rn we will use in the subsequent **while** loop. The loop compares each of the rows in the matrix cw (*i.e.*, each of the codewords) with the received vector r. The **norm** command that appears in the loop counts the number of positions in which each row of cw differs from r. If a row is found that differs from r in fewer than m positions, the variable fc is assigned the value 1. This terminates the loop, leaving m as the number of errors in r, and rn as the number of the row in cw that differs from r in fewer than m positions. If no codeword is found that differs from r in fewer than m positions, the loop ends when the rows of cw are exhausted, leaving m with its initial value of 4.

```
> m := 4:
> fc := 0:
> rn := 0:
> while (fc <> 1) and (rn < rowdim(cw)) do
>       rn := rn + 1;
>       if norm(row(cw, rn) - r, 1) < m then
>           m := norm(row(cw, rn) - r, 1):
>           fc := 1:
>       fi:
> od:
```

After we execute the preceding commands we can enter the following command to see if r contains a correctable error.

> m;

3

This value for m indicates that r contains three errors, and hence is correctable. The following command shows that the codeword that differs from r in three positions is the 22nd row of cw.

```
> rn;
```

$$22$$

We can view this codeword by entering the following command.

```
> evalm(row(cw, rn));
```

$$[0, 1, 1, 0, 1, 0, 0, 1, 0, 1, 1, 0, 1, 0, 0, 1]$$

And we can see the positions in r that contain errors by entering the following command.

```
> map(x -> x mod 2, evalm(row(cw, rn) - r));
```

$$[1, 1, 0, 0, 0, 0, 0, 0, 0, 0, 0, 0, 0, 0, 0, 1]$$

3.5 Linear Codes

As shown, Hadamard and Reed-Muller codes are easy to construct and can have significant error-correction capabilities. However, because Hadamard and Reed-Muller codes do not generally form vector spaces, they are not ideal for situations in which a very large number of codewords is needed. Because Hadamard and Reed-Muller codes do not generally form vector spaces, error correction in these codes must consist of comparing received vectors with each of the codewords one by one. This is the scheme we used to correct the received vector in Section 3.4. While this error correction scheme poses no real problems in small codes like the one constructed in Section 3.4, it would not be an efficient way to correct errors in a code with a very large number of codewords. In this section we discuss a method for constructing codes that form vector spaces. We then discuss some error correction schemes for these codes.

Recall that a code that forms a vector space is called a linear code. We will describe linear codes by the parameters $[n, k]$ if the codewords in the code are of length n positions and the code forms a vector space of dimension k. In this section we discuss linear codes constructed using *generator* matrices. Specifically, let $W = Z_2^k$ and $V = Z_2^n$ with $k < n$, and let G be a $k \times n$ matrix over Z_2 of full row rank. Then $C = \{v \in V \mid v = wG \text{ for some } w \in W\}$ is a subspace of V of dimension k. Hence, the vectors in C form the codewords in an $[n, k]$ linear code in V with 2^k codewords. The matrix G is called a generator matrix for C.

Example 3.4 Let $W = Z_2^2 = \{(00), (10), (01), (11)\}$, and choose the following generator matrix G.

$$G = \begin{bmatrix} 1 & 1 & 1 & 1 & 0 & 0 & 0 & 0 & 1 & 1 & 1 \\ 0 & 0 & 0 & 0 & 1 & 1 & 1 & 1 & 1 & 1 & 1 \end{bmatrix}$$

Then $C = \{(00000000000), (11110000111), (00001111111), (11111111000)\}$ is the resulting $[11, 2]$ linear code.   ■

Note that the code C in Example 3.4 has minimum distance $d = 7$. Hence, C is 3-error correcting, whereas errors cannot be corrected in $W = Z_2^2$. Of course, the vectors in C are longer than the vectors in W. Consequently, it would take more "effort" to transmit the vectors in C. However, the ability to correct up to 3 transmission errors in C should be much more valuable than the extra effort required to transmit the vectors. Furthermore, W can still be used for encoding and decoding actual messages or information. The general idea we can take is that messages or information can be encoded in W and then converted to C before being transmitted. Received vectors can then be corrected in C (if necessary) and converted back to W to be decoded. Note that in order for this process to be valid, we must be able to convert between W and C uniquely. But this is precisely why we required G to have full row rank. Since G has full row rank, G has a right inverse, say B. Therefore, $w \in W$ can be retrieved uniquely from $wG \in C$ by $w = wGB$.

We now consider the problem of detecting errors in received vectors that occur from codewords in linear codes constructed using generator matrices. Because linear codes are vector spaces, there are techniques for identifying received vectors as codewords in linear codes that are generally much more efficient than comparing the received vectors with each of the codewords one by one. For a linear code C constructed from W using generator matrix G of size $k \times n$, consider an $(n - k) \times n$ matrix H of full row rank over Z_2 with $HG^t = 0$. Since $HG^t = 0$, then $HG^t w^t = 0$ for all $w \in W$. Hence, $H(wG)^t = 0$ for all $w \in W$, or, equivalently, $Hc^t = 0$ for all $c \in C$. And since H has full row rank, it can be shown that $Hc^t = 0$ if and only if $c \in C$. Thus, H can be used to identify codewords in C. The matrix H is called a *parity check* matrix for C.

To determine a parity check matrix H from a generator matrix G, note that $HG^t = 0$ implies $GH^t = 0$, so the columns of H^t, that is, the rows of H, are in the null space of G. Thus, to determine H from G, we must only find a basis for the null space of G and place these basis vectors as rows in H. In practice, when constructing a linear code, it is often convenient to begin with a parity check matrix rather than a generator matrix. But since $HG^t = 0$, then G can be determined from H in the same way H can be determined from G. That is, G can be determined from H by finding a basis for the null space of H and placing these basis vectors as rows in G.

Example 3.5 Consider a linear code C with the following parity check matrix H.

$$H = \begin{bmatrix} 0 & 0 & 0 & 1 & 1 & 1 & 1 \\ 0 & 1 & 1 & 0 & 0 & 1 & 1 \\ 1 & 0 & 1 & 0 & 1 & 0 & 1 \end{bmatrix}$$

To construct a generator matrix G for C, we find a basis for the null space of H by considering H as the coefficient matrix for the following system of 3 homogeneous equations in 7 unknowns.

$$x_1 + x_3 + x_5 + x_7 = 0$$
$$x_2 + x_3 + x_6 + x_7 = 0$$
$$x_4 + x_5 + x_6 + x_7 = 0$$

By solving these equations for x_1, x_2, and x_4 in terms of the others, we can find a basis for the null space of H by setting each of x_3, x_5, x_6, and x_7 equal to one while setting the others equal to zero. For example, setting $x_5 = 1$ and $x_3 = x_6 = x_7 = 0$ gives $x_1 = x_4 = 1$ and $x_2 = 0$. This yields the basis vector (1001100). This vector and the other three basis vectors constructed similarly form the rows in the following generator matrix G.

$$G = \begin{bmatrix} 1 & 1 & 1 & 0 & 0 & 0 & 0 \\ 1 & 0 & 0 & 1 & 1 & 0 & 0 \\ 0 & 1 & 0 & 1 & 0 & 1 & 0 \\ 1 & 1 & 0 & 1 & 0 & 0 & 1 \end{bmatrix}$$

To construct the codewords in C, we would take $W = Z_2^4$ and form wG for all $w \in W$. The resulting code is a $[7, 4]$ linear code with 16 codewords called a *Hamming* code. ∎

The code in Example 3.5 is called a Hamming code because of the form of the parity check matrix H. Note that the columns of the matrix H in Example 3.5 are the numbers $1, 2, \ldots, 7$ in order expressed in binary. For example, the 6th column of H is $[1, 1, 0]^t$, whose entries are the coefficients in the expression $6 = 1 \cdot (2^2) + 1 \cdot (2^1) + 0 \cdot (2^0)$. In general, to construct a Hamming code, we place the binary expressions of the numbers $1, 2, \ldots, 2^m - 1$ for some integer $m > 1$ in order as columns in a parity check matrix H of size $m \times (2^m - 1)$. The reason we stop at a number of the form $2^m - 1$ is so that the columns of H will form all nonzero vectors of length m over Z_2. The importance of this is for error correction and will be addressed later. From H, we determine a generator matrix G of size

$(2^m - 1 - m) \times (2^m - 1)$ over Z_2 by finding a basis for the null space of H over Z_2. We then construct the codewords in the code by forming wG for all vectors w of length $2^m - 1 - m$ over Z_2.

Example 3.6 The following is the parity check matrix H for the $[15, 11]$ Hamming code.

$$H = \begin{bmatrix} 0 & 0 & 0 & 0 & 0 & 0 & 0 & 1 & 1 & 1 & 1 & 1 & 1 & 1 & 1 \\ 0 & 0 & 0 & 1 & 1 & 1 & 1 & 0 & 0 & 0 & 0 & 1 & 1 & 1 & 1 \\ 0 & 1 & 1 & 0 & 0 & 1 & 1 & 0 & 0 & 1 & 1 & 0 & 0 & 1 & 1 \\ 1 & 0 & 1 & 0 & 1 & 0 & 1 & 0 & 1 & 0 & 1 & 0 & 1 & 0 & 1 \end{bmatrix}$$

∎

All Hamming codes are one-error correcting (see Corollary 3.8) and perfect (see Written Exercise 11). Recall that a code C in Z_2^n is said to be perfect if every vector in Z_2^n is correctable in C. The fact that Hamming codes are perfect is a consequence of the discussion immediately preceding Theorem 3.1 regarding the number of correctable vectors in a t-error correcting code in Z_2^n. For example, because the $[7, 4]$ Hamming code C is one-error correcting, the number of vectors in Z_2^7 that are correctable in C is

$$|C| \cdot \left[\binom{n}{0} + \cdots + \binom{n}{t} \right] = 16 \cdot \left[\binom{7}{0} + \binom{7}{1} \right] = 128.$$

But there are only $2^7 = 128$ vectors in Z_2^7. Thus, every vector in Z_2^7 is correctable in the $[7, 4]$ Hamming code. Hence, the $[7, 4]$ Hamming code is perfect. The general result is Written Exercise 11.

We have now seen an effective method for detecting errors in received vectors that occur from codewords in linear codes constructed using generator matrices. Specifically, for a linear code C with parity check matrix H, $Hc^t = 0$ if and only if $c \in C$. We now consider the problem of correcting errors in received vectors that occur from codewords in these codes. Let C be a linear code in Z_2^n with parity check matrix H. Suppose $c \in C$ is transmitted and we receive the vector $r \in Z_2^n$. Then $r = c + e$ for some error vector $e \in Z_2^n$ that contains ones in the positions where r and c differ and zeros elsewhere. Note that $Hr^t = Hc^t + He^t = He^t$, so we can determine He^t by computing Hr^t. If we can then find e from He^t, we can form the corrected codeword as $c = r + e$.

Consider again the Hamming codes. Because they are one-error correcting and perfect, the only error vectors we must consider with Hamming codes are the vectors e_i that contain all zeros except a single one in the ith

position. Suppose a codeword in the $[2^m - 1, 2^m - 1 - m]$ Hamming code C is transmitted and we receive the vector $r \in Z_2^{2^m-1}$. If $r \notin C$, then since the columns in the parity check matrix H for C form all nonzero vectors of length m over Z_2, Hr^t will be one of the columns of H. Suppose Hr^t is the jth column of H. Since the jth column of H is also He_j^t, then $Hr^t = He_j^t$. Thus, the error in r is e_j. Note also that the jth column of H is the binary expression of the number j. Hence, if Hr^t is the binary expression of the number j, then the error in r is e_j.

Example 3.7 Suppose a codeword in the $[7, 4]$ Hamming code C is transmitted and we receive the vector $r = (1011001)$. Then with the parity check matrix H for C in Example 3.5, $Hr^t = (001)^t$ is the first column of H. Thus, the error in r is $e_1 = (1000000)$, and we correct r to the codeword $c = r + e_1 = (0011001) \in C$. ∎

Example 3.8 Suppose a codeword in the $[15, 11]$ Hamming code C is transmitted and we receive the vector $r = (101011100111000)$. Then with the parity check matrix H for C in Example 3.6, $Hr^t = (1011)^t$ is the 11th column of H. Thus, the error in r is $e_{11} = (000000000010000)$, and we correct r to the codeword $c = r + e_{11} = (101011100101000) \in C$. ∎

As mentioned, Hamming codes are one-error correcting. We now consider the problem of determining the number of errors that are guaranteed to be correctable in more general linear codes constructed using generator matrices. We discussed in Section 3.1 that we can determine the number of errors that are guaranteed to be correctable in a code by finding the minimum distance of the code. Specifically, in a code with minimum distance d, we are guaranteed to be able to uniquely correct t errors for any $t < \frac{d}{2}$. In a code with a very large number of codewords, it would not be efficient to find the minimum distance of the code by actually computing the Hamming distance between each pair of codewords. However, because linear codes are vector spaces, there are techniques for determining the minimum distance that are generally much more efficient than computing the Hamming distance between each pair of codewords. The following Theorems 3.6 and 3.7 provide such techniques.

For a codeword x in a linear code constructed using a generator matrix, we define the *Hamming weight* $w(x)$ to be the number of ones in x. That is, $w(x) = d(x, 0)$, the Hamming distance between x and the zero vector.

Theorem 3.6 *Let C be a linear code constructed using a generator matrix, and suppose $w = \min\{w(x) \mid x \in C,\ x \neq 0\}$. Then $w = d(C)$.*

Proof. Since $w = w(c) = d(c, 0)$ for some $c \in C$, it must be the case that $d(C) \leq w$. But $d(C) = d(x, y) = w(x - y)$ for some $x, y \in C$. Since C is a vector space, then $x - y \in C$. Hence, $w \leq d(C)$.   ∎

Theorem 3.7 *Let C be a linear code with parity check matrix H, and let s be the minimum number of linearly dependent columns in H. Then $s = d(C)$.*

Proof. Let $w = \min\{w(x) \mid x \in C,\ x \neq 0\}$, and suppose $C_{i_1}, \ldots, C_{i_s}$ are linearly dependent columns in H. Then

$$a_1 C_{i_1} + \cdots + a_s C_{i_s} = 0$$

for some nonzero $a_1, \ldots, a_s$. Let x be a vector of appropriate length (the number of columns in H) with a_j in position i_j for $j = 1, \ldots, s$ and zeros elsewhere. Then $Hx^t = 0$. Hence, $x \in C$. Thus, $s \geq w = d(C)$. Conversely, let $y \in C$ with $w(y) = d(C)$, and let $i_1, \ldots, i_d$ be the positions in y that are nonzero. Then $0 = Hy^t = C_{i_1} + \cdots + C_{i_d}$, so columns $C_{i_1}, \ldots, C_{i_d}$ are linearly dependent. Thus, $s \leq d(C)$.   ∎

The fact that Hamming codes are one-error correcting can be shown as a corollary to Theorem 3.7. We show this next.

Corollary 3.8 *Let C be a Hamming code. Then C is one-error correcting.*

Proof. Note that the first three columns in the parity check matrix H for C are linearly dependent. Also, no two columns in H are linearly dependent since either they would be equal or one would be the zero vector. By Theorem 3.7, $d(C) = 3$. Thus, C is one-error correcting.   ∎

We have already shown how errors can be corrected in Hamming codes. We now consider error correction in more general linear codes constructed using generator matrices.

Let C be a t-error correcting linear code in Z_2^n. A subset S of Z_2^n is called a *coset* of C if any two vectors in S differ by an element in C. Suppose $c \in C$ is transmitted and we receive the vector $r \in Z_2^n$ with $r = c + e$ for some nonzero error vector e. Since r and e differ by an element in C, then r and e are in the same coset of C. Hence, if r contains t or fewer errors, we can find the error vector e that corresponds to r by finding the unique vector with the fewest ones in the coset that contains r. In a code with a very large number of codewords, it would not be practical to construct all of the elements in the cosets. The following theorem yields an equivalence on vectors in the same coset for such codes.

Theorem 3.9 *Let C be a linear code with parity check matrix H. Then u, v are in the same coset of C if and only if $Hu^t = Hv^t$.*

Proof. Exercise.   ∎

Theorem 3.9 states that each coset S of a linear code with parity check matrix H can be uniquely identified by Hu^t for any $u \in S$. We will call Hu^t the *syndrome* of u. Suppose a codeword c in a t-error correcting linear code C in Z_2^n is transmitted and we receive the vector $r \in Z_2^n$ with $r = c + e$ for some nonzero error vector e. If r contains t or fewer errors, then we can find e by finding the unique vector with the same syndrome as r that contains t or fewer ones. And if r contains more than t errors, then the syndrome of r will not match the syndromes of any of the vectors in Z_2^n that contain t or fewer ones. When a coset contains a unique vector with the fewest number of ones, we will call this vector the *coset leader*. Hence, for a t-error correcting linear code, each vector that contains t or fewer ones must be a coset leader.

Example 3.9 Let $W = Z_2^2 = \{(00), (10), (01), (11)\}$, and choose the following generator matrix G.

$$G = \begin{bmatrix} 1 & 1 & 1 & 0 & 0 \\ 0 & 0 & 1 & 1 & 1 \end{bmatrix}$$

Then $C = \{(00000), (11100), (00111), (11011)\}$ is the resulting $[5, 2]$ linear code. It can easily be verified that the following matrix H is a parity check matrix for C.

$$H = \begin{bmatrix} 1 & 1 & 0 & 0 & 0 \\ 1 & 0 & 1 & 1 & 0 \\ 1 & 0 & 1 & 0 & 1 \end{bmatrix}$$

It can also easily be verified that C is one-error correcting. Therefore, the only cosets leaders in Z_2^5 for C will be the zero vector and the five vectors in Z_2^5 that contain a single one. The following table shows these coset leaders and their syndromes.

Coset Leader	Syndrome
(00000)	$(000)^t$
(10000)	$(111)^t$
(01000)	$(100)^t$
(00100)	$(011)^t$
(00010)	$(010)^t$
(00001)	$(001)^t$

Suppose a codeword $c \in C$ is transmitted and we receive the vector $r_1 = (00011) \in Z_2^5$. To correct this vector, we compute $Hr_1^t = (011)^t$. Since the coset leader (00100) also has this syndrome, then the error in r is $e = (00100)$. Hence, we would correct r_1 to $c = r_1 + e = (00111)$. Note that because each coset for this code contains 4 vectors, only 24 of the 32 vectors in Z_2^5 are in cosets that have coset leaders. For example, suppose a codeword in C is transmitted and we receive the vector $r_2 = (01001)$. To correct this vector we compute $Hr_2^t = (101)^t$. But none of the coset leaders for C also have this syndrome, so r_2 is not in a coset with a coset leader. Thus, r_2 cannot be corrected.   ∎

3.6 Hamming Codes with Maple

In this section we show how Maple can be used to construct codewords and correct errors in the [15, 11] Hamming code.

We begin by constructing the parity check matrix H for the code. We first enter the length $m = 4$ of the binary vectors that form the columns in the parity check matrix.

```
> m := 4:
```

Recall that the columns of H are binary expressions of length m for the numbers $1, 2, \ldots, 2^m - 1$. We can obtain the binary expression of a number in Maple by using the **convert** command. For example, we can obtain the binary expression of the number 4 by entering the following command.

```
> cb := convert(4, base, 2);
```

$$cb := [\, 0,\, 0,\, 1\,]$$

The entries in the preceding vector are the coefficients in the expression $4 = 0 \cdot (2^0) + 0 \cdot (2^1) + 1 \cdot (2^2)$. Note that this vector contains only three positions, whereas for the columns in H we want binary vectors of length $m = 4$ positions. That is, to be placed as the 4th column in H, we would want the number 4 to be converted to the binary vector $[0, 0, 1, 0]$. Furthermore, note that the binary digits in this vector are the reverse of how they should be expressed in the 4th column of H. To be directly placed as the 4th column in H, the number 4 should be converted to the binary vector $[0, 1, 0, 0]$. We can use the following commands to take care of these problems. After first including the Maple **linalg** package, we define the vector **bv** of length $m = 4$ containing all zeros. Then in the subsequent **for** loop we place the binary digits from **cb** in appropriate order in the vector **bv**.

```
> with(linalg):
> bv := vector(m, 0):
> for i from 1 to vectdim(cb) do
>      bv[m-i+1] := cb[i]:
> od:
> evalm(bv);
```

$$[0, 1, 0, 0]$$

We now construct the parity check matrix H for the $[15, 11]$ Hamming code by placing properly ordered binary representations of length $m = 4$ for the numbers $1, 2, \ldots, 2^m - 1$ as columns in H. To do this, we first create an empty list **H**. We then use a Maple **for** loop to build the parity check matrix column by column in **H**. The **op** command that appears in the loop allows new columns to be attached to **H** with the **augment** command.

```
> H := []:
> for j from 1 to 2^m-1 do
>      cb := convert(j, base, 2):
>      bv := vector(m, 0):
>      for i from 1 to vectdim(cb) do
>           bv[m-i+1] := cb[i]:
>      od:
>      H := augment(op(H), bv):
> od:
> evalm(H);
```

$$\begin{bmatrix} 0 & 0 & 0 & 0 & 0 & 0 & 0 & 1 & 1 & 1 & 1 & 1 & 1 & 1 & 1 \\ 0 & 0 & 0 & 1 & 1 & 1 & 1 & 0 & 0 & 0 & 0 & 1 & 1 & 1 & 1 \\ 0 & 1 & 1 & 0 & 0 & 1 & 1 & 0 & 0 & 1 & 1 & 0 & 0 & 1 & 1 \\ 1 & 0 & 1 & 0 & 1 & 0 & 1 & 0 & 1 & 0 & 1 & 0 & 1 & 0 & 1 \end{bmatrix}$$

Next, we construct a generator matrix G for the $[15, 11]$ Hamming code by finding a basis for the null space of **H** over Z_2 and placing these basis vectors as rows in G. To do this, we first find a basis for the null space of **H** using the Maple **Nullspace** command as follows.

```
> nH := Nullspace(H) mod 2;
```

$$nH := \{[0, 1, 0, 1, 0, 1, 0, 0, 0, 0, 0, 0, 0, 0, 0],$$
$$[0, 0, 0, 1, 0, 0, 0, 1, 0, 0, 0, 1, 0, 0, 0],$$
$$[1, 1, 0, 1, 0, 0, 1, 0, 0, 0, 0, 0, 0, 0, 0],$$
$$[1, 1, 0, 0, 0, 0, 0, 1, 0, 0, 1, 0, 0, 0, 0],$$

$$[1, 0, 0, 1, 0, 0, 0, 1, 0, 0, 0, 0, 1, 0, 0],$$
$$[1, 1, 0, 1, 0, 0, 0, 1, 0, 0, 0, 0, 0, 0, 1],$$
$$[1, 0, 0, 0, 0, 0, 0, 1, 1, 0, 0, 0, 0, 0, 0],$$
$$[1, 1, 1, 0, 0, 0, 0, 0, 0, 0, 0, 0, 0, 0, 0],$$
$$[1, 0, 0, 1, 1, 0, 0, 0, 0, 0, 0, 0, 0, 0, 0],$$
$$[0, 1, 0, 1, 0, 0, 0, 1, 0, 0, 0, 0, 0, 1, 0],$$
$$[0, 1, 0, 0, 0, 0, 0, 1, 0, 1, 0, 0, 0, 0, 0]\}$$

The preceding output is a basis for the null space of H expressed as rows in a set. Because Maple places a default ordering on the vectors in this set (although not necessarily in the order in which they are displayed), each basis vector can be retrieved by entering a command like the following.

> nH[2];

$$[0, 1, 0, 0, 0, 0, 0, 1, 0, 1, 0, 0, 0, 0, 0]$$

The following command returns the number of vectors in the set nH.

> nops(nH);

$$11$$

We now form a generator matrix G for the [15, 11] Hamming code by placing the vectors in the set nH as rows in G. In the following **for** loop we build the generator matrix row by row in G. Note that we attach new rows to G using the Maple **stackmatrix** command.[3]

```
> G := []:
> for i from 1 to nops(nH) do
>       G := stackmatrix(op(G), nH[i]):
> od:
> evalm(G);
```

$$
\begin{bmatrix}
0 & 1 & 0 & 1 & 0 & 0 & 0 & 1 & 0 & 0 & 0 & 0 & 0 & 1 & 0 \\
0 & 1 & 0 & 0 & 0 & 0 & 0 & 1 & 0 & 1 & 0 & 0 & 0 & 0 & 0 \\
1 & 0 & 0 & 0 & 0 & 0 & 0 & 1 & 1 & 0 & 0 & 0 & 0 & 0 & 0 \\
1 & 1 & 1 & 0 & 0 & 0 & 0 & 0 & 0 & 0 & 0 & 0 & 0 & 0 & 0 \\
1 & 0 & 0 & 1 & 1 & 0 & 0 & 0 & 0 & 0 & 0 & 0 & 0 & 0 & 0 \\
1 & 1 & 0 & 1 & 0 & 0 & 1 & 0 & 0 & 0 & 0 & 0 & 0 & 0 & 0 \\
1 & 1 & 0 & 0 & 0 & 0 & 0 & 1 & 0 & 0 & 1 & 0 & 0 & 0 & 0 \\
1 & 0 & 0 & 1 & 0 & 0 & 0 & 1 & 0 & 0 & 0 & 0 & 1 & 0 & 0 \\
1 & 1 & 0 & 1 & 0 & 0 & 0 & 1 & 0 & 0 & 0 & 0 & 0 & 0 & 1 \\
0 & 0 & 0 & 1 & 0 & 0 & 0 & 1 & 0 & 0 & 0 & 1 & 0 & 0 & 0 \\
0 & 1 & 0 & 1 & 0 & 1 & 0 & 0 & 0 & 0 & 0 & 0 & 0 & 0 & 0
\end{bmatrix}
$$

---

[3] See footnote p. 51 regarding the Maple **stackmatrix** command.

Recall that the codewords in a linear code constructed using a generator matrix G are all vectors of the form wG where w is a vector over Z_2 of appropriate length. To see the length of the vectors w for the $[15, 11]$ Hamming code, we can enter the following command, which returns the number of rows in G.

> rowdim(G);

$$11$$

Hence, the vectors w for the $[15, 11]$ Hamming code should contain 11 positions. For example, consider the following vector w.

> w := vector([1, 0, 1, 1, 1, 0, 1, 1, 1, 1, 0]):

In the next command we form the codeword wG that results from w. Note that we use the **map** command to reduce the result over Z_2. Note also that we use the Maple **&*** command for matrix multiplication.

> c := map(x -> x mod 2, evalm(w &* G));

$$c := [0, 0, 1, 1, 1, 0, 0, 0, 1, 0, 1, 1, 1, 1, 1]$$

We now show how Maple can be used to correct errors in the $[15, 11]$ Hamming code. Suppose a codeword in the $[15, 11]$ Hamming code is transmitted and we receive the following vector r.

> r := vector([0, 1, 1, 1, 1, 1, 0, 1, 1, 1, 1, 1, 1, 1, 0]):

To determine if an error exists in this received vector, we compute the syndrome of r as follows.

> syn := map(x -> x mod 2, evalm(H &* r));

$$syn := [1, 0, 0, 1]$$

Because this syndrome is nonzero, we know r is not a codeword in the $[15, 11]$ Hamming code. Recall that to correct the error in r we must only find the column in H that matches this syndrome. The number of this column in H will be the same as the number of the position in r that contains an error. We can use the following commands to find the column in H that matches this syndrome. We first assign the parameters fc and cn that we will use in the subsequent **while** loop. The loop compares each of the columns in H with the syndrome of r. When a match is found, the variable fc is assigned the value 1. This terminates the loop, leaving cn as the column number where the match occurred. The **col** command that appears in the loop allows us to access each of the columns of H. The Maple **equal** command is a logical statement that returns *true* if its parameters are equal, and *false* if not.

> fc := 0:
> cn := 0:

```
> while (fc <> 1) and (cn < 2^m-1) do
>       cn := cn + 1;
>       if equal(col(H, cn), syn) = true then
>             fc := 1;
>       fi:
> od:
```

After we execute the preceding commands, we can enter the following command to see the position in r that contains an error.

```
> cn;
```

$$9$$

This value for cn indicates that the error in r is in the 9th position. To correct this error, we first define the following vector e of length $2^m - 1$ containing all zeros.

```
> e := vector(2^m-1, 0);
```

$$e := [0, 0, 0, 0, 0, 0, 0, 0, 0, 0, 0, 0, 0, 0, 0]$$

Next, we define the entry in position cn of e to be equal to 1.

```
> e[cn] := 1:
```

We can then see the error vector that corresponds to r as follows.

```
> evalm(e);
```

$$[0, 0, 0, 0, 0, 0, 0, 0, 1, 0, 0, 0, 0, 0, 0]$$

And we can see the corrected codeword as follows.

```
> map(x -> x mod 2, evalm(r + e));
```

$$[0, 1, 1, 1, 1, 1, 0, 1, 0, 1, 1, 1, 1, 1, 0]$$

Written Exercises

1. Using Theorem 3.1, find the maximum number of errors that are guaranteed to be correctable in a code of length 7 with 4 codewords. Then, using Theorem 3.3, show that it is not possible to construct a code of length 7 with 4 codewords that is guaranteed to correct this maximum number of errors.

2. Construct a $(15, 8)$ code with 16 codewords. What is the maximum number of errors that are guaranteed to be correctable in this code?

3. Construct an $(8, 4)$ code with 16 codewords. What is the maximum number of errors that are guaranteed to be correctable in this code?

4. Is it possible to construct a $[6, 2]$ linear code that is 2-error correcting? State how you know. (Hint: See Theorem 3.1.)

5. Find a generator matrix for a 2-error correcting linear code with 4 codewords. Also, construct a parity check matrix for the code.

6. Let C be the $[7, 4]$ Hamming code.

   (a) Construct the codewords in C.

   (b) Correct the following received vectors in C: $r_1 = (0011101)$, $r_2 = (0100101)$.

   (c) Make a list of the coset leaders for C and their syndromes.

7. For the code in Example 3.9, which of the vectors $r_1 = (11101)$, $r_2 = (01011)$, and $r_3 = (10101)$ can be corrected using the coset method? Correct those that can be corrected. For the one(s) that cannot be corrected, explain why.

8. Let $W = \{(00), (01), (10), (11)\}$, and choose the following generator matrix G.
$$G = \begin{bmatrix} 1 & 1 & 0 & 0 & 1 & 1 \\ 0 & 0 & 1 & 1 & 1 & 1 \end{bmatrix}$$

   (a) Construct the linear code C that results from G and W. How many errors are guaranteed to be correctable in this code?

   (b) Make a list of the coset leaders for C and their syndromes for all coset leaders that contain a single one or all zeros. Why are the remaining cosets irrelevant for error correction?

   (c) Which of the vectors $r_1 = (100011)$, $r_2 = (001100)$, and $r_3 = (111100)$ can be corrected in C using the coset method? Correct those that can be corrected. For the one(s) that cannot be corrected, explain why.

9. Prove Corollary 3.4.

10. Prove Theorem 3.5.

11. Show that the $[2^m - 1, 2^m - 1 - m]$ Hamming codes are perfect.

12. Prove Theorem 3.9.

13. An *equivalence relation* is a relation $\sim$ on a set A that satisfies
    (i) $a \sim a$
    (ii) $a \sim b \Rightarrow b \sim a$
    (iii) $a \sim b$ and $b \sim c \Rightarrow a \sim c$
    for every $a, b, c \in A$. Let C be a linear code in Z_2^n. Define a relation $\sim$ on V by $x \sim y$ if x and y are in the same coset of C. Show that $\sim$ is an equivalence relation on V.

14. A *metric space* is a set M with a real-valued function $d(\cdot, \cdot)$ on $M \times M$ that satisfies
    (i) $d(x, y) \geq 0$, and $d(x, y) = 0$ if and only if $x = y$
    (ii) $d(x, y) = d(y, x)$
    (iii) $d(x, z) \leq d(x, y) + d(y, z)$
    for every $x, y, z \in M$. Prove or disprove: a code C in Z_2^n with the Hamming distance function $d(\cdot, \cdot)$ is a metric space.

Maple Exercises

1. We mentioned in Section 3.3 that the $(32, 16)$ Reed-Muller code was used in the Mariner 9 space probe when it returned photographs of Mars to Earth in 1972.

   (a) Construct the codewords in the $(32, 16)$ Reed-Muller code. How many errors are guaranteed to be correctable in this code?

   (b) Correct the following received vector r in the $(32, 16)$ Reed-Muller code,

   $$r = (11100101011010011110101101101001)$$

2. Let C be the $[31, 26]$ Hamming code.

   (a) Construct the parity check matrix and a generator matrix for C.

   (b) Construct the codeword in C that results from the following vector w.

   $$w = (10110101110110111110111000)$$

   (c) Correct the following received vector r in C.

   $$r = (1101011100110110110101011110111)$$

Chapter 4

BCH Codes

The most useful codes we discussed in Chapter 3 were Hamming codes because they are linear and perfect. However, Hamming codes are not ideal for situations in which the occurrence of more than one error in a single codeword is likely. Recall that Hamming codes are only one-error correcting. If more than one error occurs during the transmission of a Hamming codeword, the received vector will not be correctable to the codeword that was sent. Furthermore, since Hamming codes are perfect, if more than one error occurs during the transmission of a Hamming codeword, the received vector will be uniquely correctable – it will just be correctable to the wrong codeword. In this chapter we discuss a class of codes called *BCH* codes that are linear and can be constructed to be multiple-error correcting. BCH codes are named for their creators Bose, Chaudhuri, and Hocquenghem.

4.1   Construction of BCH Codes

One way that BCH codes differ from the codes we discussed in Chapter 3 is that BCH codewords are polynomials rather than vectors. To construct a BCH code, we begin by letting $f(x) = x^m - 1 \in Z_2[x]$ for some positive integer m. Then $R = Z_2[x]/(f(x))$ is a ring that can be represented by all polynomials in $Z_2[x]$ of degree less than m. Suppose $g(x) \in Z_2[x]$ divides $f(x)$. Then $C = \{$multiples of $g(x)$ in $Z_2[x]$ of degree less than $m\}$ is a vector space in R with dimension $m - \deg g(x)$. Hence, the polynomials in C form codewords in an $[m, m - \deg g(x)]$ linear code in R with $2^{m - \deg g(x)}$ codewords. The polynomial $g(x)$ is called a *generator polynomial* for the code. We consider the codewords in this code to have length m positions because we view each term in a polynomial codeword as a codeword position. A codeword $c(x) \in Z_2[x]$ with m terms can naturally be expressed as a unique vector in Z_2^m by listing the coefficients of $c(x)$ in order (including coefficients of zero). In this book we will assume BCH codewords are transmitted this way with increasing powers of x.

Example 4.1 Let $f(x) = x^7 - 1$ and $g(x) = x^3 + x + 1$ in $Z_2[x]$. Then the code C of multiples of $g(x)$ in $Z_2[x]$ of degree less than 7 has basis $\{ x^3 + x + 1, \, x^4 + x^2 + x, \, x^5 + x^3 + x^2, \, x^6 + x^4 + x^3 \}$. Hence, C is a $[7, 4]$ code with 16 codewords consisting of all linear combinations of these basis polynomials in $Z_2[x]$. In this code, we will assume that the codeword $x^5 + x^4 + x^3 + x$ would be transmitted as the vector $0 + 1x + 0x^2 + 1x^3 + 1x^4 + 1x^5 + 0x^6 = (0101110) \in Z_2^7$. ∎

For a code constructed as described above to be a BCH code, the generator polynomial $g(x)$ must be chosen as follows. Let $a_1, a_2, \ldots, a_s$ for $s < m$ be roots of $f(x)$ with minimum polynomials $m_1(x), m_2(x), \ldots, m_s(x)$ in $Z_2[x]$, respectively, and let $g(x)$ be the least common multiple of the polynomials $m_i(x)$ in $Z_2[x]$. Note that $g(x)$ divides $f(x)$, so $g(x)$ can be used as the generator polynomial for a code. Choosing $g(x)$ in this manner is useful because of how it allows errors to be corrected in the resulting code. We will discuss BCH error correction in Section 4.2. Actually, choosing a generator polynomial as just described still does not necessarily yield a BCH code. For the resulting code to be a BCH code, the values of m and the roots a_i must be chosen in a special way. We describe this next.

Let $m = 2^n - 1$ for some positive integer n, and let $f(x) = x^m - 1$ in $Z_2[x]$. Suppose $p(x)$ is a primitive polynomial of degree n in $Z_2[x]$. Then $Z_2[x]/(p(x))$ is a field of order 2^n whose nonzero elements are generated by the field element x. For reasons that will become apparent when we begin discussing Reed-Solomon codes in Chapter 5, we will denote the element x in this field by a. Then, for the roots a_i described in the previous paragraph, we let $a_i = a^i$ for $i = 1, \ldots, s$. Choosing the a_i in this manner is useful because of how it allows the generator polynomial $g(x)$ to be determined. The polynomials $m_i(x)$ described in the previous paragraph are then the minimum polynomials of a^i for $i = 1, \ldots, s$. Thus, we can determine $g(x)$ by forming the product that includes a single factor of each unique $m_i(x)$. As a consequence of Lagrange's Theorem (Theorem 1.4), a^i will be a root of $f(x)$ for all i. Hence, $g(x)$ will divide $f(x)$.

Because BCH codewords are in $Z_2[x]$, some of the computations that are necessary for constructing BCH codes can be done very easily. Specifically, note that $(x_1 + x_2 + \cdots + x_r)^2 = x_1^2 + x_2^2 + \cdots + x_r^2$ over Z_2 since all cross terms will contain a factor of 2. Therefore, for a polynomial

$h(x) = x^{i_1} + x^{i_2} + \cdots + x^{i_r} \in Z_2[x]$, it follows that

$$h(a^2) = (a^2)^{i_1} + (a^2)^{i_2} + \cdots + (a^2)^{i_r} = (a^{i_1} + a^{i_2} + \cdots + a^{i_r})^2 = h(a)^2.$$

Similarly, it can be seen that $h(a^{2k}) = h(a^k)^2$ for any positive integer k. Thus, for example, $h(a^{12}) = h(a^6)^2 = h(a^3)^4$. The utility of this will be clear in the following examples.

Example 4.2 Let $f(x) = x^7 - 1$, and choose the primitive polynomial $p(x) = x^3 + x + 1$ in $Z_2[x]$. Then for the element $a = x$ in the field $Z_2[x]/(p(x))$ of order 8, we list the field elements that correspond to the first seven powers of a in the following table.

Power	Field Element
a^1	a
a^2	a^2
a^3	$a + 1$
a^4	$a^2 + a$
a^5	$a^2 + a + 1$
a^6	$a^2 + 1$
a^7	1

Let C be the BCH code that results from considering the first four powers of a. To determine the generator polynomial $g(x)$ for C, we must find the minimum polynomials $m_1(x)$, $m_2(x)$, $m_3(x)$, and $m_4(x)$. But since $p(x)$ is primitive and $a = x$, it follows that $p(a) = 0$. Furthermore, $p(a^2) = p(a)^2 = 0$ and $p(a^4) = p(a)^4 = 0$ since $p(x) \in Z_2[x]$. Thus, $m_1(x) = m_2(x) = m_4(x) = p(x)$. Now, since a^3 is a root of $f(x)$, the minimum polynomial $m_3(x)$ of a^3 must be one of the irreducible factors of $x^7 - 1 = (x^3 + x + 1)(x^3 + x^2 + 1)(x + 1)$. (This factorization can be obtained by using the Maple **Factor** command as illustrated in Section 4.3.1.) By substituting a^3 into each of these irreducible factors, we can find that $x^3 + x^2 + 1$ is equal to zero when evaluated at a^3. Hence, $m_3(x) = x^3 + x^2 + 1$. Thus, $g(x) = m_1(x)m_3(x) = x^6 + x^5 + x^4 + x^3 + x^2 + x + 1$. The code that results from this generator polynomial is a $[7, 1]$ BCH code with basis $\{g(x)\}$ and two codewords. ∎

Example 4.3 Let $f(x) = x^{15} - 1$, and choose the primitive polynomial $p(x) = x^4 + x + 1$ in $Z_2[x]$. Then for the element $a = x$ in the field $Z_2[x]/(p(x))$ of order 16, we list the field elements that correspond to the first 15 powers of a in the following table.

Power	Field Element	Power	Field Element
a^1	a	a^9	$a^3 + a$
a^2	a^2	a^{10}	$a^2 + a + 1$
a^3	a^3	a^{11}	$a^3 + a^2 + a$
a^4	$a + 1$	a^{12}	$a^3 + a^2 + a + 1$
a^5	$a^2 + a$	a^{13}	$a^3 + a^2 + 1$
a^6	$a^3 + a^2$	a^{14}	$a^3 + 1$
a^7	$a^3 + a + 1$	a^{15}	1
a^8	$a^2 + 1$		

Let C be the BCH code that results from considering the first six powers of a. To determine the generator polynomial $g(x)$ for C, we must find the minimum polynomials $m_1(x), m_2(x), \ldots, m_6(x)$. But $p(a) = 0$, and hence $p(a^2) = p(a^4) = 0$. Thus, $m_1(x) = m_2(x) = m_4(x) = p(x)$. Also, since a^3 and a^5 are roots of $f(x)$, then $m_3(x)$ and $m_5(x)$ are irreducible factors of
$$x^{15} - 1 = (x+1)(x^2 + x + 1)(x^4 + x + 1)(x^4 + x^3 + 1)(x^4 + x^3 + x^2 + x + 1).$$
By substituting a^3 and a^5 into each of these irreducible factors, we can find that $m_3(x) = x^4 + x^3 + x^2 + x + 1$ and $m_5(x) = x^2 + x + 1$. Furthermore, $m_3(a^6) = m_3(a^3)^2 = 0$, and hence $m_6(x) = m_3(x)$. Thus, $g(x) = m_1(x)m_3(x)m_5(x) = x^{10} + x^8 + x^5 + x^4 + x^2 + x + 1$. The code that results from this generator polynomial is a $[15, 5]$ BCH code with basis $\{g(x), xg(x), x^2 g(x), x^3 g(x), x^4 g(x)\}$ and $2^5 = 32$ codewords. ∎

Although BCH codes are not as easy to construct as Hadamard or Reed-Muller codes, BCH codes are linear while Hadamard and Reed-Muller codes are not. Also, unlike Hamming codes, BCH codes can be constructed to be multiple-error correcting. Specifically, in the next section we will show that a BCH code that results from considering the first $2t$ powers of a is t-error correcting. For example, since in Example 4.3 we considered the first six powers of a, the resulting BCH code is 3-error correcting. Also, since in Example 4.2 we considered the first four powers of a, the resulting BCH code is 2-error correcting. We discuss a scheme for correcting errors in BCH codewords next.

4.2 Error Correction in BCH Codes

As mentioned in Section 4.1, the generator polynomial for a BCH code is chosen in a special way because of how it allows errors to be corrected in the code. Before discussing the BCH error correction scheme, we first note the following theorem.

Theorem 4.1 *Let C be a BCH code that results from a primitive polynomial of degree n by considering the first s powers of a, and suppose $c(x) \in Z_2[x]$ has degree less than $2^n - 1$. Then $c(x) \in C$ if and only if $c(a^i) = 0$ for $i = 1, \ldots, s$.*

Proof. Let $m_i(x)$ be the minimum polynomial of a^i in $Z_2[x]$ for $i = 1, \ldots, s$, and let $g(x)$ be the least common multiple of the polynomials $m_i(x)$ in $Z_2[x]$. If $c(x) \in C$, then $c(x) = g(x)h(x)$ for some $h(x) \in Z_2[x]$. Thus, $c(a^i) = g(a^i)h(a^i) = 0\,h(a^i) = 0$ for $i = 1, \ldots, s$. Conversely, if $c(a^i) = 0$ for $i = 1, \ldots, s$, then $m_i(x)$ divides $c(x)$ for $i = 1, \ldots, s$. Hence, $g(x)$ divides $c(x)$, and $c(x) \in C$. ∎

We now outline the BCH error correction scheme. Let C be a BCH code that results from a primitive polynomial of degree n by considering the first $2t$ powers of a. We will show in Theorem 4.2 that C is then t-error correcting. Suppose $c(x) \in C$ is transmitted and we receive the polynomial $r(x) \neq c(x)$ in $Z_2[x]$ of degree less than $2^n - 1$. Then $r(x) = c(x) + e(x)$ for some nonzero error polynomial $e(x)$ in $Z_2[x]$ of degree less than $2^n - 1$. To correct $r(x)$, we must only determine $e(x)$, for we could then compute $c(x) = r(x) + e(x)$. But note that Theorem 4.1 implies $r(a^i) = e(a^i)$ for $i = 1, \ldots, 2t$. Thus, by knowing $r(x)$, we also know some information about $e(x)$. We will call the values of $r(a^i)$ the *syndromes* of $r(x)$. Now, suppose

$$e(x) = x^{m_1} + x^{m_2} + \cdots + x^{m_p}$$

for some integer error positions $m_1 < m_2 < \cdots < m_p$ with $p \leq t$ and $m_p < 2^n - 1$. To find these error positions, we begin by computing the first $2t$ syndromes of $r(x)$. We will denote these syndromes as follows by $r_1, r_2, \ldots, r_{2t}$.

$$
\begin{aligned}
r_1 &= r(a) &&= e(a) &&= a^{m_1} + a^{m_2} + \cdots + a^{m_p} \\
r_2 &= r(a^2) &&= e(a^2) &&= (a^2)^{m_1} + (a^2)^{m_2} + \cdots + (a^2)^{m_p} \\
&\ \ \vdots \\
r_{2t} &= r(a^{2t}) &&= e(a^{2t}) &&= (a^{2t})^{m_1} + (a^{2t})^{m_2} + \cdots + (a^{2t})^{m_p}
\end{aligned}
$$

Next, we introduce the following polynomial $E(z)$, which we will call an *error locator polynomial*.

$$
\begin{aligned}
E(z) &= (z - a^{m_1})(z - a^{m_2}) \cdots (z - a^{m_p}) \\
&= z^p + \sigma_1 z^{p-1} + \cdots + \sigma_p
\end{aligned}
$$

We call $E(z)$ an *error locator polynomial* because the roots of $E(z)$ show the error positions in $r(x)$. Our eventual goal will be to determine these roots. Before doing this, we must first find the coefficients $\sigma_1, \sigma_2, \ldots, \sigma_p$ of $E(z)$. These coefficients are the elementary symmetric functions in $a^{m_1}, a^{m_2}, \ldots, a^{m_p}$. That is,

$$\sigma_1 = \sum_{1 \le i \le p} a^{m_i}$$

$$\sigma_2 = \sum_{1 \le i < j \le p} a^{m_i} a^{m_j}$$

$$\vdots$$

$$\sigma_p = a^{m_1} \cdots a^{m_p}.$$

Note that if we evaluate $E(a^{m_j})$ for all $1 \le j \le p$ and multiply each result by $(a^{m_j})^i$ for any $1 \le i \le p$, since $E(a^{m_j}) = 0$ for all $1 \le j \le p$, we obtain the following system of equations for $1 \le i \le p$.

$$0 = (a^{m_1})^i [(a^{m_1})^p + \sigma_1 (a^{m_1})^{(p-1)} + \cdots + \sigma_p]$$
$$0 = (a^{m_2})^i [(a^{m_2})^p + \sigma_1 (a^{m_2})^{(p-1)} + \cdots + \sigma_p]$$

$$\vdots$$

$$0 = (a^{m_p})^i [(a^{m_p})^p + \sigma_1 (a^{m_p})^{(p-1)} + \cdots + \sigma_p]$$

By distributing the $(a^{m_j})^i$ in the preceding equations and summing the results, we obtain the following equation for $1 \le i \le p$.

$$0 = r_{i+p} + \sigma_1 r_{i+p-1} + \sigma_2 r_{i+p-2} + \cdots + \sigma_p r_i$$

Since this holds for $1 \le i \le p$, this yields a system of p linear equations in the p unknowns $\sigma_1, \ldots, \sigma_p$ that are equivalent to the following single matrix equation.

$$\begin{bmatrix} r_1 & \cdots & r_p \\ \vdots & & \vdots \\ r_p & \cdots & r_{2p-1} \end{bmatrix} \begin{bmatrix} \sigma_p \\ \vdots \\ \sigma_1 \end{bmatrix} = \begin{bmatrix} r_{p+1} \\ \vdots \\ r_{2p} \end{bmatrix} \tag{4.1}$$

If the $p \times p$ coefficient matrix in (4.1) is nonsingular, then we can solve (4.1) uniquely for $\sigma_1, \ldots, \sigma_p$. After we find $\sigma_1, \ldots, \sigma_p$, we can then form the error locator polynomial $E(z)$ and determine $a^{m_1}, \ldots, a^{m_p}$ by trial and error as the roots of $E(z)$. This reveals the error positions $m_1, \ldots, m_p$ in $r(x)$.

We will now look at two examples of the BCH error correction scheme in the code that results from the generator polynomial in Example 4.3. Since we will generally not know the number of errors in a received polynomial before attempting to correct it, we will begin the BCH error correction scheme in a t-error correcting BCH code by assuming that the received polynomial $r(x)$ contains the maximum number t of correctable errors and using the first $2t$ syndromes of $r(x)$. If $r(x)$ does not contain exactly t errors, then the $t \times t$ coefficient matrix in (4.1) will not be nonsingular. In this case, we can simply reduce the number of assumed errors to $t-1$ and repeat the error correction procedure using only the first $2t-2$ syndromes of $r(x)$. If $r(x)$ also does not contain exactly $t-1$ errors (*i.e.*, if the $(t-1) \times (t-1)$ coefficient matrix in (4.1) is also not nonsingular), we can continue to repeat the procedure, each time reducing the number of assumed errors by one and using twice as many syndromes as the number of assumed errors until the coefficient matrix in (4.1) is nonsingular. If the error in $r(x)$ is not correctable, then the coefficient matrix in (4.1) will not be nonsingular for any number of assumed errors between 1 and t.

Example 4.4 Let C be the BCH code in Example 4.3 that results from the generator polynomial $g(x) = x^{10} + x^8 + x^5 + x^4 + x^2 + x + 1$. Suppose a codeword in C is transmitted as a vector in Z_2^{15} and we receive the vector $r = (101111110010000) \in Z_2^{15}$. Note first that this vector converts to the polynomial $r(x) = 1 + x^2 + x^3 + x^4 + x^5 + x^6 + x^7 + x^{10} \in Z_2[x]$. It can easily be verified that $g(x)$ does not divide $r(x)$. Hence, $r(x) \notin C$. Since C is 3-error correcting, to correct $r(x)$ we begin by computing the first six syndromes of $r(x)$. Using the table of powers of a and corresponding field elements in Example 4.3, we can compute these syndromes as follows.

$$
\begin{aligned}
r_1 &= r(a) \\
&= 1 + a^2 + a^3 + a^4 + a^5 + a^6 + a^7 + a^{10} \\
&= 1 + a^2 + a^3 + a + 1 + a^2 + a + a^3 + a^2 + a^3 + a + 1 + a^2 + a + 1 \\
&= a^3
\end{aligned}
$$

$$
\begin{aligned}
r_3 &= r(a^3) \\
&= 1 + a^6 + a^9 + a^{12} + a^{15} + a^{18} + a^{21} + a^{30} \\
&= 1 + a^6 + a^9 + a^{12} + 1 + a^3 + a^6 + 1 \\
&= \cdots \\
&= a^6
\end{aligned}
$$

$$
\begin{aligned}
r_5 &= r(a^5) \\
&= 1 + a^{10} + a^{15} + a^{20} + a^{25} + a^{30} + a^{35} + a^{50} \\
&= \cdots \\
&= a^{10}
\end{aligned}
$$

Since $r(x) \in Z_2[x]$, we can find the remaining syndromes as follows.

$$
\begin{aligned}
r_2 &= r(a^2) &= (r(a))^2 &= (a^3)^2 &= a^6 \\
r_4 &= r(a^4) &= (r(a))^4 &= (a^3)^4 &= a^{12} \\
r_6 &= r(a^6) &= (r(a^3))^2 &= (a^6)^2 &= a^{12}
\end{aligned}
$$

Now, assuming $r(x)$ contains three errors, we must find σ_1, σ_2, and σ_3 that satisfy the following equation.

$$\begin{bmatrix} a^3 & a^6 & a^6 \\ a^6 & a^6 & a^{12} \\ a^6 & a^{12} & a^{10} \end{bmatrix} \begin{bmatrix} \sigma_3 \\ \sigma_2 \\ \sigma_1 \end{bmatrix} = \begin{bmatrix} a^{12} \\ a^{10} \\ a^{12} \end{bmatrix} \tag{4.2}$$

It can be verified that the determinant of the 3×3 coefficient matrix in (4.2) is a^{12}. Hence, this coefficient matrix is nonsingular and $r(x)$ contains three errors. We can use Cramer's rule to determine σ_1, σ_2, and σ_3. For example, since

$$\begin{vmatrix} a^{12} & a^6 & a^6 \\ a^{10} & a^6 & a^{12} \\ a^{12} & a^{12} & a^{10} \end{vmatrix} = a^{28} + a^{30} + a^{28} + a^{24} + a^{36} + a^{26}$$

$$= 1 + a^9 + a^6 + a^{11}$$
$$= 1 + a^3 + a + a^3 + a^2 + a^3 + a^2 + a$$
$$= a^3 + 1$$
$$= a^{14},$$

Cramer's rule yields

$$\sigma_3 = \frac{a^{14}}{a^{12}} = a^2.$$

Similarly, since

$$\begin{vmatrix} a^3 & a^{12} & a^6 \\ a^6 & a^{10} & a^{12} \\ a^6 & a^{12} & a^{10} \end{vmatrix} = \cdots = a^{10} \quad \text{and} \quad \begin{vmatrix} a^3 & a^6 & a^{12} \\ a^6 & a^6 & a^{10} \\ a^6 & a^{12} & a^{12} \end{vmatrix} = \cdots = 1,$$

Cramer's rule yields

$$\sigma_2 = \frac{a^{10}}{a^{12}} = a^{13} \quad \text{and} \quad \sigma_1 = \frac{1}{a^{12}} = a^3.$$

The resulting error locator polynomial is $E(z) = z^3 + a^3 z^2 + a^{13} z + a^2$. By evaluating $E(z)$ at successive powers of a, we can find that the roots of $E(z)$ are 1, a^5, and a^{12}. Hence, the error in $r(x)$ is $e(x) = 1 + x^5 + x^{12}$. Thus, we correct $r(x)$ to the following codeword $c(x)$.

$$c(x) = r(x) + e(x) = x^2 + x^3 + x^4 + x^6 + x^7 + x^{10} + x^{12}$$

It can easily be verified that this polynomial $c(x)$ is a multiple of $g(x)$.

Suppose another codeword in C is transmitted and we receive the vector $r = (100100010011010) \in Z_2^{15}$. This vector converts to the polynomial

$r(x) = 1 + x^3 + x^7 + x^{10} + x^{11} + x^{13} \in Z_2[x]$. It can easily be verified that $r(x)$ is not a multiple of $g(x)$. Hence, $r(x) \notin C$. To correct $r(x)$ we begin by computing the first six syndromes of $r(x)$. These syndromes can be determined as in the first part of this example and are as follows.

$$
\begin{aligned}
r_1 &= a^5 \\
r_2 &= a^{10} \\
r_3 &= a^2 \\
r_4 &= a^5 \\
r_5 &= 1 \\
r_6 &= a^4
\end{aligned}
$$

Now, assuming $r(x)$ contains three errors, we must find σ_1, σ_2, and σ_3 that satisfy the following equation.

$$
\begin{bmatrix}
a^5 & a^{10} & a^2 \\
a^{10} & a^2 & a^5 \\
a^2 & a^5 & 1
\end{bmatrix}
\begin{bmatrix}
\sigma_3 \\
\sigma_2 \\
\sigma_1
\end{bmatrix}
=
\begin{bmatrix}
a^5 \\
1 \\
a^4
\end{bmatrix}
\tag{4.3}
$$

However, it can be verified that the determinant of the 3×3 coefficient matrix in (4.3) is 0. Hence, this coefficient matrix is not nonsingular and $r(x)$ does not contain three errors. Thus, we assume $r(x)$ contains only two errors and use only the first four syndromes of $r(x)$. Assuming $r(x)$ contains only two errors, we must find σ_1 and σ_2 that satisfy the following equation.

$$
\begin{bmatrix}
a^5 & a^{10} \\
a^{10} & a^2
\end{bmatrix}
\begin{bmatrix}
\sigma_2 \\
\sigma_1
\end{bmatrix}
=
\begin{bmatrix}
a^2 \\
a^5
\end{bmatrix}
\tag{4.4}
$$

It can be verified that the determinant of the 2×2 coefficient matrix in (4.4) is a^{13}, so this coefficient matrix is nonsingular and $r(x)$ contains two errors. We can again use Cramer's rule to determine σ_1 and σ_2. Specifically, since

$$
\begin{vmatrix}
a^2 & a^{10} \\
a^5 & a^2
\end{vmatrix}
= \cdots = a
\qquad \text{and} \qquad
\begin{vmatrix}
a^5 & a^2 \\
a^{10} & a^5
\end{vmatrix}
= \cdots = a^3,
$$

Cramer's rule yields

$$
\sigma_2 = \frac{a}{a^{13}} = a^3
\qquad \text{and} \qquad
\sigma_1 = \frac{a^3}{a^{13}} = a^5.
$$

The resulting error locator polynomial is $E(z) = z^2 + a^5 z + a^3$. The roots of $E(z)$ can be determined as a and a^2. Hence, the error in $r(x)$ is $e(x) = x + x^2$ and we correct $r(x)$ to the following codeword $c(x)$.

$$
c(x) = r(x) + e(x) = 1 + x + x^2 + x^3 + x^7 + x^{10} + x^{11} + x^{13}
$$

It can easily be verified that this polynomial $c(x)$ is a multiple of $g(x)$. ∎

We close this section by proving a fundamental result we have mentioned regarding BCH codes.

Theorem 4.2 *Let C be a BCH code that results from considering the first $2t$ powers of a. Then C is t-error correcting.*

Proof. Suppose C results from a primitive polynomial of degree n, and let $m = 2^n - 1$. Consider the following matrix H.

$$H = \begin{bmatrix} 1 & a & a^2 & \cdots & a^{m-1} \\ 1 & a^2 & (a^2)^2 & \cdots & (a^2)^{m-1} \\ 1 & a^3 & (a^3)^2 & \cdots & (a^3)^{m-1} \\ \vdots & \vdots & \vdots & & \vdots \\ 1 & a^{2t} & (a^{2t})^2 & \cdots & (a^{2t})^{m-1} \end{bmatrix}$$

Note that for a polynomial $r(x) = b_0 + b_1 x + \cdots + b_{m-1} x^{m-1} \in Z_2[x]$, if we let $b = (b_0, b_1, \ldots, b_{m-1})$, then $Hb^t = (r_1, r_2, \ldots, r_{2t})^t$. Hence, $r(x) \in C$ if and only if Hb^t is the zero vector. Thus, H can serve as a parity check matrix for C.

We will now show that the minimum number of linearly dependent columns in H is $2t + 1$. We first show that any $2t$ columns in H must be linearly independent. Choose integers $0 \le j_1 < j_2 < \cdots < j_{2t} < m$. Then the columns in H in these positions form the following $2t \times 2t$ matrix.

$$\begin{bmatrix} a^{j_1} & a^{j_2} & \cdots & a^{j_{2t}} \\ (a^2)^{j_1} & (a^2)^{j_2} & \cdots & (a^2)^{j_{2t}} \\ \vdots & \vdots & & \vdots \\ (a^{2t})^{j_1} & (a^{2t})^{j_2} & \cdots & (a^{2t})^{j_{2t}} \end{bmatrix}$$

The determinant of this matrix can be expressed as

$$\begin{vmatrix} 1 & 1 & \cdots & 1 \\ a^{j_1} & a^{j_2} & & a^{j_{2t}} \\ \vdots & \vdots & & \vdots \\ (a^{j_1})^{2t-1} & (a^{j_2})^{2t-1} & \cdots & (a^{j_{2t}})^{2t-1} \end{vmatrix} \cdot a^{j_1} a^{j_2} \cdots a^{j_{2t}},$$

which is nonzero because it is the determinant of a Vandermonde matrix with distinct columns. Thus, any $2t$ columns in H are linearly independent. Also, since H has $2t$ rows, then we know that any $2t + 1$ columns in H must be linearly dependent. Therefore, the minimum number of linearly dependent columns in H is $2t + 1$. By Theorem 3.7, we know then that the minimum distance of C is $2t + 1$. Hence, C is t-error correcting. ∎

4.3 BCH Codes with Maple

In this section, we show how Maple can be used to construct the BCH generator polynomial in Example 4.3, and correct the two received polynomials in Example 4.4.

We begin by including the Maple **linalg** package and entering the primitive polynomial $p(x) = x^4 + x + 1 \in Z_2[x]$ used to construct the code.

```
> with(linalg):
> p := x -> x^4 + x + 1:
> Primitive(p(x)) mod 2;
```
$$true$$

Next, we use the Maple **degree** function to assign the number of elements in the underlying field as the variable **fs**, and use the Maple **vector** function to create a vector in which to store the field elements.

```
> fs := 2^(degree(p(x)));
```
$$fs := 16$$

```
> field := vector(fs);
```
$$field := \text{array}(\, 1..16, [\,]\,)$$

By entering the following commands, we generate and store the field elements in the vector **field**. Since for BCH codes we denote the field element x by a, we use the parameters **a** and **p(a)** in the following **Powmod** command.

```
> for i from 1 to fs-1 do
>      field[i] := Powmod(a, i, p(a), a) mod 2:
> od:
> field[fs] := 0:
> evalm(field);
```
$$\begin{aligned}
&[\, a,\, a^2,\, a^3,\, a+1,\, a^2+a,\, a^3+a^2,\, a^3+a+1,\, a^2+1,\, a^3+a, \\
&\quad a^2+a+1,\, a^3+a^2+a,\, a^3+a^2+a+1,\, a^3+a^2+1,\, a^3+1, \\
&\quad 1,\, 0\,]
\end{aligned}$$

Because working with BCH codes requires frequent conversions between polynomial field elements and powers of a, it will be very useful for us to establish an association between the polynomial field elements and corresponding powers of a for this field. We will establish this association in a table. We first use the Maple **table** command to create a table.

```
> ftable := table():
```

Then, by entering the following commands we establish an association between the polynomial field elements and corresponding powers of a for this field in `ftable`. Note the bracket `[ ]` syntax for accessing the field elements and the table entries.

```
> for i from 1 to fs-1 do
>       ftable[ field[i] ] := a^i:
> od:
> ftable[ field[fs] ] := 0:
```

We can view the entries in `ftable` by entering the following **print** command. For the sake of space (because Maple displays this table as a vertical list) we have removed this output.

```
> print(ftable);
```

The following command illustrates how `ftable` can be used. Specifically, by entering the following command we can access the power of a that corresponds to the polynomial field element $a^3 + a^2$.

```
> ftable[a^3 + a^2];
```

$$a^6$$

4.3.1   Construction of the Generator Polynomial

We now show how Maple can be used to construct the generator polynomial in Example 4.3. We first define the polynomial $f(x) = x^{15} - 1$ of which each power of a is a root.

```
> f := x -> x^(fs-1) - 1;
```

$$f := x \rightarrow x^{(fs-1)} - 1$$

Next, we use the Maple **Factor** command to find the irreducible factors of $f(x)$ in $Z_2[x]$.

```
> factf := Factor(f(x)) mod 2;
```

$$factf := (\, x^4 + x^3 + x^2 + x + 1 \,)(\, x^4 + x + 1 \,)(\, x^2 + x + 1 \,)$$
$$(\, x + 1 \,)(\, x^4 + x^3 + 1 \,)$$

To construct the generator polynomial in Example 4.3, we will need to access the factors of $f(x)$ separately. We can do this by using the Maple **op** command. For example, we can use the following command to assign the third factor in the expression `factf` to the variable `f3`.

```
> f3 := op(3, factf);
```

$$f3 := x^2 + x + 1$$

We can then use the Maple **unapply** command as follows to convert `f3` into a function that can be evaluated in the usual manner.

```
> f3 := unapply(f3, x);
```

$$f3 := x \rightarrow x^2 + x + 1$$

```
> f3(a^6);
```

$$a^{12} + a^6 + 1$$

The following Maple **Rem** command returns the polynomial field element that corresponds to the preceding output.

```
> Rem(f3(a^6), p(a), a) mod 2;
```

$$a$$

And the following command returns the number of factors in `factf`.

```
> nops(factf);
```

$$5$$

We now find the minimum polynomials that will be the factors in the generator polynomial. We first assign the number $t = 3$ of errors the code is to be able to correct. In the subsequent loops we find and display the minimum polynomials of $a, a^2, \ldots, a^{2t}$. In these commands, the outer loop spans the powers a^i of a, while the inner loop evaluates each factor of `factf` at a^i. Since each power of a is a root of `factf`, each power of a will be a root of an irreducible factor of `factf`. The factor of which each power of a is a root will be the minimum polynomial of that power of a. The **if** and **print** statements that appear in these commands cause the correct minimum polynomial to be displayed. The **break** statement causes the inner loop to terminate when the correct minimum polynomial is found.

```
> t := 3:
> for i from 1 to 2*t do
>       for j from 1 to nops(factf) do
>           fj := op(j, factf):
>           fj := unapply(fj, x):
>           if Rem(fj(a^i), p(a), a) mod 2 = 0 then
>               print(a^i, '   is a root of   ', fj(x)):
>               break:
>           fi:
>       od:
> od:
```

$$a, \quad \text{is a root of} \quad , x^4 + x + 1$$

$$a^2, \quad \text{is a root of} \quad , x^4 + x + 1$$

$$a^3, \quad \text{is a root of} \quad , x^4 + x^3 + x^2 + x + 1$$

$$a^4, \quad \text{is a root of} \quad , x^4 + x + 1$$

$$a^5, \quad \text{is a root of} \quad , x^2 + x + 1$$

$$a^6, \quad \text{is a root of} \quad , x^4 + x^3 + x^2 + x + 1$$

Next, we define one factor of each of the three unique minimum polynomials in the preceding output.

```
> m1 := x -> x^4 + x + 1:
> m3 := x -> x^4 + x^3 + x^2 + x + 1:
> m5 := x -> x^2 + x + 1:
```

We can then define the generator polynomial $g(x)$ as the product of these three factors.

```
> g := m1(x) * m3(x) * m5(x);
```

$$g := (x^4 + x + 1)(x^4 + x^3 + x^2 + x + 1)(x^2 + x + 1)$$

Finally, we convert $g(x)$ into a function that can be evaluated in the usual manner as follows.

```
> g := unapply(g, x);
```

$$g := x \longrightarrow (x^4 + x + 1)(x^4 + x^3 + x^2 + x + 1)(x^2 + x + 1)$$

4.3.2   Error Correction

We now show how Maple can be used to correct the received polynomials in Example 4.4. Consider first the following received polynomial $r(x)$.

```
> r := x -> 1 + x^2 + x^3 + x^4 + x^5 + x^6 + x^7 + x^10:
```

Recall that to correct $r(x)$ we begin by computing the first $2t$ syndromes of $r(x)$. Before doing this, we first create a vector **syn** of length $2t$ in which to store the syndromes.

```
> syn := vector(2*t);
```

$$syn := \text{array}(1..6, [\,])$$

By entering the following commands we generate and store the first $2t$ syndromes of $r(x)$ in syn.

```
> for i from 1 to 2*t do
>       syn[i] := Rem(r(a^i), p(a), a) mod 2:
>       syn[i] := ftable[ syn[i] ];
> od:
> evalm(syn);
```

$$[a^3, a^6, a^6, a^{12}, a^{10}, a^{12}]$$

We can then access particular syndromes of $r(x)$ from the vector syn. For example, we can access the 5th syndrome of $r(x)$ by entering the following command.

```
> syn[5];
```

$$a^{10}$$

Next, we define the 3×3 coefficient matrix from (4.2) as follows.

```
> A := matrix( [ [syn[1], syn[2], syn[3]], [syn[2], syn[3],
> syn[4]], [syn[3], syn[4], syn[5]] ] );
```

$$A := \begin{bmatrix} a^3 & a^6 & a^6 \\ a^6 & a^6 & a^{12} \\ a^6 & a^{12} & a^{10} \end{bmatrix}$$

And next we define the vector from the right-hand side of (4.2).

```
> b := vector( [syn[4], syn[5], syn[6]] );
```

$$b := [a^{12}, a^{10}, a^{12}]$$

We can use the Maple det function to find the determinant of A as follows.

```
> d := det(A);
```

$$d := a^{19} - a^{27} - a^{22} + 2\,a^{24} - a^{18}$$

The next command returns the field element that corresponds to the determinant of A.

```
> d := Rem(det(A), p(a), a) mod 2;
```

$$d := a^3 + a^2 + a + 1$$

And the next command returns the power of a that corresponds to the determinant of **A**.

```
> d := ftable[d];
```

$$d := a^{12}$$

Since this determinant is nonzero, we know $r(x)$ contains three errors. As in Example 4.4, we will use Cramer's rule to determine σ_1, σ_2, and σ_3 from (4.2). To construct the matrices required for Cramer's rule, we will use the Maple **col** function for choosing a column from a matrix. For example, the following command returns the second column from **A**.

```
> col(A, 2);
```

$$[\, a^6, a^6, a^{12} \,]$$

We can easily use the Maple **col** and **augment** functions to construct the matrices required for Cramer's rule. For example, the following command constructs a new matrix **A1** from **A** by replacing the first column of **A** with the vector **b**. This matrix is necessary for Cramer's rule in computing σ_3.

```
> A1 := augment(b, col(A,2), col(A,3));
```

$$A1 := \begin{bmatrix} a^{12} & a^6 & a^6 \\ a^{10} & a^6 & a^{12} \\ a^{12} & a^{12} & a^{10} \end{bmatrix}$$

The following command returns the determinant of **A1** as a power of a.

```
> dA1 := ftable[ Rem(det(A1), p(a), a) mod 2 ];
```

$$dA1 := a^{14}$$

Hence, by Cramer's Rule, we can compute σ_3 as follows.

```
> sigma3 := dA1/d;
```

$$\sigma3 := a^2$$

Similarly, by Cramer's Rule, we can determine σ_2 as follows.

```
> A2 := augment(col(A,1), b, col(A,3));
```

$$A2 := \begin{bmatrix} a^3 & a^{12} & a^6 \\ a^6 & a^{10} & a^{12} \\ a^6 & a^{12} & a^{10} \end{bmatrix}$$

```
> dA2 := ftable[ Rem( det(A2), p(a), a) mod 2 ];
```

$$dA2 := a^{10}$$

```
> sigma2 := dA2/d;
```

$$\sigma2 := \frac{1}{a^2}$$

Because this expression for σ_2 in not in the exact form we want (as a positive power of a), we enter the following command.

```
> sigma2 := sigma2 * a^(fs-1);
```

$$\sigma2 := a^{13}$$

Finally, by Cramer's Rule we can determine σ_1 as follows.

```
> A3 := augment(col(A,1), col(A,2), b);
```

$$A3 := \begin{bmatrix} a^3 & a^6 & a^{12} \\ a^6 & a^6 & a^{10} \\ a^6 & a^{12} & a^{12} \end{bmatrix}$$

```
> dA3 := ftable[ Rem(det(A3), p(a), a) mod 2 ];
```

$$dA3 := a^{15}$$

```
> sigma1 := dA3/d;
```

$$\sigma1 := a^3$$

Next, we define the resulting error locator polynomial.

```
> EL := z^3 + sigma1*z^2 + sigma2*z + sigma3;
```

$$EL := z^3 + a^3 z^2 + a^{13} z + a^2$$

```
> EL := unapply(EL, z);
```

$$EL := z \rightarrow z^3 + a^3 z^2 + a^{13} z + a^2$$

By entering the following commands we find the roots of this error locator polynomial by trial and error.

```
> for i from 1 to fs-1 do
>       if Rem(EL(a^i), p(a), a) mod 2 = 0 then
>           print(a^i, '  is a root of EL(z) =  ', EL(z));
>       fi:
> od:
```

$$a^5, \quad \text{is a root of } EL(z) = \quad , z^3 + a^3 z^2 + a^{13} z + a^2$$

$$a^{12}, \quad \text{is a root of } EL(z) = \quad , z^3 + a^3 z^2 + a^{13} z + a^2$$

$$a^{15}, \quad \text{is a root of } EL(z) = \quad , z^3 + a^3 z^2 + a^{13} z + a^2$$

Recall, $a^{15} = 1$. Hence, the roots of this error locator polynomial are 1, a^5, and a^{12}. Thus, the following polynomial is the error $e(x) = 1 + x^5 + x^{12}$ in the received polynomial $r(x)$.

```
> e := x -> 1 + x^5 + x^12:
```

The next command returns the corrected codeword.

```
> c := (r(x) + e(x)) mod 2;
```

$$c := x^2 + x^3 + x^4 + x^6 + x^7 + x^{10} + x^{12}$$

And the next command verifies that $c(x)$ is a multiple of $g(x)$.

```
> (Factor(c) mod 2)/g(x);
```

$$x^2$$

We now consider the following polynomial $r(x)$, which is the second received polynomial from Example 4.4.

```
> r := x -> 1 + x^3 + x^7 + x^10 + x^11 + x^13:
```

To correct this received polynomial, we begin as follows by computing and storing the first $2t$ syndromes of $r(x)$ in the vector **syn**.

```
> for i from 1 to 2*t do
>       syn[i] := Rem(r(a^i), p(a), a) mod 2:
>       syn[i] := ftable[ syn[i] ];
> od:
> evalm(syn);
```

$$[a^5, a^{10}, a^2, a^5, a^{15}, a^4]$$

Next, we define the coefficient matrix and vector from (4.3).

```
> A := matrix( [ [syn[1], syn[2], syn[3]], [syn[2], syn[3],
> syn[4]], [syn[3], syn[4], syn[5]] ] );
```

$$A := \begin{bmatrix} a^5 & a^{10} & a^2 \\ a^{10} & a^2 & a^5 \\ a^2 & a^5 & a^{15} \end{bmatrix}$$

```
> b := vector( [syn[4], syn[5], syn[6]] );
```

$$b := [\, a^5,\ a^{15},\ a^4 \,]$$

We find the determinant of the coefficient matrix from (4.3) as follows.

```
> d := Rem(det(A), p(a), a) mod 2;
```

$$d := 0$$

Since this determinant is zero, we know $r(x)$ does not contain exactly three errors. Thus, we assume $r(x)$ contains only two errors and define the coefficient matrix and vector from (4.4). To create the coefficient matrix, we first use the Maple **delrows** command to delete the last row from A.

```
> A := delrows(A, 3..3);
```

$$A := \begin{bmatrix} a^5 & a^{10} & a^2 \\ a^{10} & a^2 & a^5 \end{bmatrix}$$

Note that the last column of this new matrix A is the vector on the right-hand side of (4.4). We define this vector next.

```
> b := col(A, 3);
```

$$b := [\, a^2,\ a^5 \,]$$

Then, by deleting the last column from the new matrix A as follows, we obtain the coefficient matrix from (4.4).

```
> A := delcols(A, 3..3);
```

$$A := \begin{bmatrix} a^5 & a^{10} \\ a^{10} & a^2 \end{bmatrix}$$

Next, we find the determinant of this 2×2 coefficient matrix.

```
> d := ftable[ Rem(det(A), p(a), a) mod 2 ];
```

$$d := a^{13}$$

Since this determinant is nonzero, we know $r(x)$ contains two errors. We again use Cramer's rule to determine σ_1 and σ_2 from (4.4).

```
> A1 := augment(b, col(A,2));
```

$$A1 := \begin{bmatrix} a^2 & a^{10} \\ a^5 & a^2 \end{bmatrix}$$

```
> dA1 := ftable[ Rem(det(A1), p(a), a) mod 2 ];
```

$$dA1 := a$$

```
> sigma2 := dA1/d;
```

$$\sigma2 := \frac{1}{a^{12}}$$

```
> sigma2 := sigma2 * a^(fs-1);
```

$$\sigma2 := a^3$$

```
> A2 := augment(col(A,1), b);
```

$$A2 := \begin{bmatrix} a^5 & a^2 \\ a^{10} & a^5 \end{bmatrix}$$

```
> dA2 := ftable[ Rem(det(A2), p(a), a) mod 2 ];
```

$$dA2 := a^3$$

```
> sigma1 := dA2/d;
```

$$\sigma1 := \frac{1}{a^{10}}$$

```
> sigma1 := sigma1 * a^(fs-1);
```

$$\sigma1 := a^5$$

Next, we define the resulting error locator polynomial.

```
> EL := z^2 + sigma1*z + sigma2;
```

$$EL := z^2 + a^5 z + a^3$$

```
> EL := unapply(EL, z);
```

$$EL := z \rightarrow z^2 + a^5 z + a^3$$

We find the roots of this error locator polynomial as follows.

```
> for i from 1 to fs-1 do
>       if Rem(EL(a^i), p(a), a) mod 2 = 0 then
>           print(a^i, '   is a root of EL(z) =  ', EL(z));
>       fi:
> od:
```

$$a, \quad \textit{is a root of } EL(z) \; = \; , z^2 + a^5 z + a^3$$

$$a^2, \quad \textit{is a root of } EL(z) \; = \; , z^2 + a^5 z + a^3$$

Thus, the following polynomial is the error $e(x) = x + x^2$ in $r(x)$.

```
> e := x -> x + x^2:
```

The next command returns the corrected codeword.

```
> c := (r(x) + e(x)) mod 2;
```

$$c := 1 + x^3 + x^7 + x^{10} + x^{11} + x^{13} + x + x^2$$

We can use the Maple **sort** command as follows to sort the terms in the preceding polynomial.

```
> sort(c);
```

$$x^{13} + x^{11} + x^{10} + x^7 + x^3 + x^2 + x + 1$$

The next command verifies that $c(x)$ is a multiple of $g(x)$.

```
> (Factor(c) mod 2)/g(x);
```

$$(x^2 + x + 1)(x + 1)$$

Written Exercises

1. Use the primitive polynomial $p(x) = x^3 + x^2 + 1 \in Z_2[x]$ to construct a generator polynomial for a one-error correcting BCH code. State the parameters $[n, k]$ for the code.

2. Correct the following received vectors in the BCH code that results from the generator polynomial in Written Exercise 1.

   (a) $r = (1101110)$
   (b) $r = (1100010)$

3. Use the primitive polynomial $p(x) = x^3 + x^2 + 1 \in Z_2[x]$ to construct a generator polynomial for a 2-error correcting BCH code. State the parameters $[n, k]$ for the code. How does this code compare to the code that results from the generator polynomial in Example 4.2?

4. Use the primitive polynomial $p(x) = x^3 + x^2 + 1 \in Z_2[x]$ to construct a generator polynomial for a 3-error correcting BCH code. State the parameters $[n, k]$ for the code. How does this code compare to the code that results from the generator polynomial in Written Exercise 3? Can you make any additional conclusions about the code that results from the generator polynomial in Written Exercise 3?

5. Use the primitive polynomial $p(x) = x^4 + x + 1 \in Z_2[x]$ to construct a generator polynomial for a $[15, 7]$ BCH code. State the number of codewords in the code and the number of errors that can be corrected in the code. How does this code compare to the code that results from the generator polynomial in Example 4.3?

6. Use the primitive polynomial $p(x) = x^4 + x^3 + 1 \in Z_2[x]$ to construct a generator polynomial for a 2-error correcting BCH code. State the code's parameters $[n, k]$ and the number of codewords in the code.

7. Use the primitive polynomial $p(x) = x^4 + x^3 + 1 \in Z_2[x]$ to construct a generator polynomial for a 3-error correcting BCH code. State the code's parameters $[n, k]$ and the number of codewords in the code.

8. Correct the following received vectors in the BCH code that results from the generator polynomial in Example 4.3.

   (a) $r = (100011011001010)$
   (b) $r = (011111010011010)$
   (c) $r = (101000011101100)$
   (d) $r = (111011001010100)$

9. The irreducible factors of $f(x) = x^{31} - 1 \in Z_2[x]$ are $(x + 1)$ and six primitive polynomials of degree 5. To construct a t-error correcting BCH code with binary codewords of length 31 positions, we begin with a primitive polynomial $p(x) \in Z_2[x]$ of degree 5, and a certain number of powers of $a = x$ in the field of order 32 resulting from $p(x)$. This determines a generator polynomial for the code, the number of basis elements in the code, and the number of codewords in the code. Complete the following table for a BCH code with binary codewords of length 31 positions.

Number of Correctable Errors	Number of Powers of a Needed	Degree of Generator Polynomial	Number of Basis Elements	Number of Codewords
3				
4				
5				
6				

Maple Exercises

1. Find a primitive polynomial of degree 5 in $Z_2[x]$, and use this polynomial to construct a generator polynomial for a BCH code. State the parameters $[n, k]$ for the code, the number of codewords in the code, and the number of errors that can be corrected in the code.

2. Use the primitive polynomial $p(x) = x^6 + x^5 + 1 \in Z_2[x]$ to construct a generator polynomial for a 3-error correcting BCH code. State the code's parameters $[n, k]$ and the number of codewords in the code.

3. Correct the following received polynomials in the BCH code that results from the generator polynomial in Maple Exercise 2.

(a) $r(x) = 1 + x + x^2 + x^9 + x^{10} + x^{13} + x^{15} + x^{16} + x^{17} + x^{18} + x^{21} + x^{22} + x^{25} + x^{26} + x^{28}$

(b) $r(x) = 1 + x + x^2 + x^3 + x^5 + x^6 + x^7 + x^8 + x^9 + x^{11} + x^{12} + x^{14} + x^{15} + x^{18} + x^{20} + x^{21} + x^{22} + x^{23}$

(c) $r(x) = x^3 + x^4 + x^5 + x^{12} + x^{14} + x^{18} + x^{19} + x^{20} + x^{21}$

Chapter 5

Reed-Solomon Codes

In this chapter we discuss a class of codes called *Reed-Solomon* codes. These codes, like BCH codes, have polynomial codewords, are linear, and can be constructed to be multiple-error correcting. However, Reed-Solomon codes are significantly more popular than BCH codes and all other types of codes because they are uniquely ideal for correcting error bursts. A received vector is said to contain an *error burst* if it contains several errors very close together. There are many situations in which transmission errors in binary vectors occur naturally in bursts. We describe one such situation, an application of a Reed-Solomon code in the Voyager 2 satellite that returned photographs to Earth of several of the planets in our solar system, in Section 5.6. Reed-Solomon codes also have numerous other applications. For example, another extensive and well-known use of Reed-Solomon codes is in the encoding of music, software, and other information on compact discs.

5.1 Construction of Reed-Solomon Codes

To construct a Reed-Solomon code, we begin as in the construction of a BCH code by choosing a primitive polynomial $p(x)$ of degree n in $Z_2[x]$ and forming the field $F = Z_2[x]/(p(x))$ of order 2^n. We will again denote the element x in this field by a. Reed-Solomon codewords, like BCH codewords, are then polynomials of degree less than $2^n - 1$. However, unlike BCH codewords which are elements in $Z_2[x]$, Reed-Solomon codewords are in $F[x]$. To construct a t-error correcting Reed-Solomon code C, we use the following generator polynomial $g(x) \in F[x]$.

$$g(x) = (x - a)(x - a^2) \cdots (x - a^{2t})$$

91

The codewords in C are then all multiples $b(x)g(x)$ of degree less than 2^n-1 with $b(x) \in F[x]$. Theorem 4.2 can easily be modified to show that C is t-error correcting. The codewords in C have length $2^n - 1$ positions and form a vector space with dimension $2^n - 1 - 2t$. We will use the notation $RS(2^n - 1, t)$ to represent a t-error correcting Reed-Solomon code with codewords of length $2^n - 1$ positions.

Example 5.1 Choose primitive polynomial $p(x) = x^4 + x + 1 \in Z_2[x]$. The nonzero elements in $F = Z_2[x]/(p(x))$ are listed in the table in Example 4.3. Using this field F, we obtain the following generator polynomial $g(x)$ for an $RS(15, 2)$ code C.

$$
\begin{aligned}
g(x) &= (x - a)(x - a^2)(x - a^3)(x - a^4) \\
&= \cdots \\
&= x^4 + a^{13}x^3 + a^6x^2 + a^3x + a^{10}
\end{aligned}
$$

To construct one of the codewords in C, consider $b(x) = a^{10}x^9 \in F[x]$. Then

$$b(x)g(x) = a^{10}x^{13} + a^8x^{12} + ax^{11} + a^{13}x^{10} + a^5x^9$$

is one of the codewords in C.   ∎

The fact that Reed-Solomon codewords are in $F[x]$ causes two problems we must address. First, unlike BCH codewords, Reed-Solomon codewords cannot be transmitted as binary vectors by simply listing the coefficients. Despite this, Reed-Solomon codewords are still transmitted as binary vectors. We will discuss transmission of Reed-Solomon codewords in Section 5.4. The other problem we must address is that of error correction, for the BCH error correction scheme cannot be used to correct errors in Reed-Solomon codewords. Actually, applying the BCH error correction scheme to a received Reed-Solomon polynomial yields the same information as when it is applied to a received BCH polynomial. Recall that the last step in the BCH error correction scheme involves finding the roots of the error locator polynomial. This reveals only the error positions in the received polynomial. However, because there are only two possible coefficients for each term in a BCH polynomial, knowledge of the error positions is all that is necessary to correct the polynomial. The BCH error correction scheme can also be used to find the error positions in a received Reed-Solomon polynomial, but because there is more than one possible coefficient for each term in a Reed-Solomon polynomial, knowledge of the error positions is generally not enough to correct the polynomial. The specific error in each error position must also be determined. Hence, we must present a new error correction scheme for correcting Reed-Solomon polynomials. We discuss this error correction scheme next.

5.2 Error Correction in Reed-Solomon Codes

Before stating the Reed-Solomon error correction scheme, we first note the following analogue to Theorem 4.1.

Theorem 5.1 *Let F be a field of order 2^n, and let C be an $RS(2^n - 1, t)$ code in $F[x]$. Suppose $c(x) \in F[x]$ has degree less than $2^n - 1$. Then $c(x) \in C$ if and only if $c(a^i) = 0$ for $i = 1, \ldots, 2t$.*

Proof. Exercise.   ∎

Theorem 5.1 is useful for error correction in Reed-Solomon codes in the same way Theorem 4.1 is useful for error correction in BCH codes. Specifically, let F be a field of order 2^n, and let C be an $RS(2^n - 1, t)$ code in $F[x]$. Suppose $c(x) \in C$ is transmitted and we receive the polynomial $r(x) \neq c(x)$ in $F[x]$ of degree less than $2^n - 1$. Then $r(x) = c(x) + e(x)$ for some nonzero error polynomial $e(x)$ in $F[x]$ of degree less than $2^n - 1$. To correct $r(x)$ we must only determine $e(x)$, for we could then compute $c(x) = r(x) + e(x)$. But note that Theorem 5.1 implies $r(a^i) = e(a^i)$ for $i = 1, \ldots, 2t$. Thus, by knowing $r(x)$, we also know some information about $e(x)$. We will again call the values of $r(a^i)$ the syndromes of $r(x)$.

We now outline the Reed-Solomon error correction scheme. As we will show, this error correction scheme is only slightly more computationally intensive than the BCH error correction scheme. However, because verification of the Reed-Solomon error correction scheme is significantly more involved than verification of the BCH error correction scheme, we will not verify the Reed-Solomon error-correction scheme in this section. In this section we will summarize and illustrate the Reed-Solomon error correction scheme. We will then verify the Reed-Solomon error correction scheme in Section 5.3.

Let F be a field of order 2^n, and let C be an $RS(2^n - 1, t)$ code in $F[x]$. Suppose $c(x) \in C$ is transmitted and we receive the polynomial $r(x) = c(x) + e(x)$ for some nonzero error polynomial $e(x)$ in $F[x]$ of degree less than $2^n - 1$. We can use the following steps to determine $e(x)$.

1. We first compute the first $2t$ syndromes of $r(x)$, which we will denote by $S_1 = r(a), S_2 = r(a^2), \ldots, S_{2t} = r(a^{2t})$, and form the following *syndrome polynomial* $S(z)$.

$$S(z) = S_1 + S_2 z + S_3 z^2 + \cdots + S_{2t} z^{2t-1}$$

(Note: Because $r(x)$ is not necessarily in $Z_2[x]$, it will not necessarily be the case that $r(a^{2k}) = r(a^k)^2$ for any integer k.)

2. Next, we construct the Euclidean algorithm table (see Section 1.6) for the polynomials $a(z) = z^{2t}$ and $b(z) = S(z)$ in $F[z]$, stopping at the first row j for which $\deg(r_j) < t$. (The **U** column may be excluded from this table.) Let $R(z) = r_j$ and $V(z) = v_j$.

3. We can then find the error positions in $r(x)$ by finding the roots of $V(z)$. Specifically, if $a^{i_1}, a^{i_2}, \ldots, a^{i_k}$ are the roots of $V(z)$, then $r(x)$ contains errors in positions $x^{-i_1}, x^{-i_2}, \ldots, x^{-i_k}$. Finally, we must find the coefficients of $e(x)$ at these error positions. Let e_{-i} be the coefficient of the x^{-i} term in $e(x)$. Then $e_{-i} = \dfrac{R(a^i)}{V'(a^i)}$.

We will illustrate this error correction scheme first in a BCH code by correcting the first received vector in Example 4.4. Although the BCH error correction scheme is not sufficient to correct errors in Reed-Solomon codewords, the Reed-Solomon error correction scheme can be used to correct errors in BCH codewords.

Example 5.2 Let C be the BCH code that results from the generator polynomial in Example 4.3. Suppose a codeword $c(x) \in C$ is transmitted and we receive the polynomial $r(x) = 1 + x^2 + x^3 + x^4 + x^5 + x^6 + x^7 + x^{10}$. In this example we will correct $r(x)$ using the Reed-Solomon error correction scheme. Since C is 3-error correcting, we begin by computing the first six syndromes of $r(x)$. These syndromes were computed in Example 4.4 and are as follows.

$$\begin{array}{rcccl}
S_1 &=& r(a) &=& a^3 \\
S_2 &=& r(a^2) &=& a^6 \\
S_3 &=& r(a^3) &=& a^6 \\
S_4 &=& r(a^4) &=& a^{12} \\
S_5 &=& r(a^5) &=& a^{10} \\
S_6 &=& r(a^6) &=& a^{12}
\end{array}$$

These syndromes yield the following syndrome polynomial $S(z)$.

$$S(z) = a^3 + a^6 z + a^6 z^2 + a^{12} z^3 + a^{10} z^4 + a^{12} z^5$$

Constructing the Euclidean algorithm table for $a(z) = z^6$ and $b(z) = S(z)$ (with numerous necessary calculations omitted), we obtain the following.

Row	Q	R	V
-1	$-$	z^6	0
0	$-$	$S(z)$	1
1	$a^3 z + a$	$a^{12} z^4 + a^{10} z^3 + z^2 + a^{10} z + a^4$	$a^3 z + a$
2	z	$a^{11} z^3 + a^7 z^2 + a^{12} z + a^3$	$a^3 z^2 + az + 1$
3	$az + a^5$	$a^4 z^2 + a^5$	$a^4 z^3 + z^2 + a^5 z + a^2$

Note that we have not included the $\mathbf{U}$ column in this table, and that we have stopped at the first row for which the degree of the entry in the $\mathbf{R}$ column is less than the number of errors that can be corrected in C. Next, we let $R(z) = a^4 z^2 + a^5$ and $V(z) = a^4 z^3 + z^2 + a^5 z + a^2$. By evaluating $V(z)$ at successive powers of a, we can find that the roots of $V(z)$ are a^0, a^3, and a^{10}. The positions in $r(x)$ that contain errors are x^0, $x^{-3} = x^{12}$, and $x^{-10} = x^5$. To determine the coefficients of these terms in the error polynomial, we first note that $V'(z) = a^4 z^2 + a^5$. We can then determine the coefficients of the terms in the error polynomial as follows.

$$e_0 = \frac{R(a^0)}{V'(a^0)} = \frac{a^4 + a^5}{a^4 + a^5} = 1$$

$$e_5 = \frac{R(a^{10})}{V'(a^{10})} = \frac{a^{24} + a^5}{a^{24} + a^5} = 1$$

$$e_{12} = \frac{R(a^3)}{V'(a^3)} = \frac{a^{10} + a^5}{a^{10} + a^5} = 1$$

Hence, the error in $r(x)$ can be expressed as the error polynomial $e(x) = 1 \cdot x^0 + 1 \cdot x^5 + 1 \cdot x^{12} = 1 + x^5 + x^{12}$. ■

Although the error correction procedure in Example 5.2 appears less involved than the procedure used to correct the same received polynomial in Example 4.4, the procedure in Example 5.2 is actually more involved due to the many calculations necessary in constructing the Euclidean algorithm table. Also, Example 5.2 shows a relatively simple example of the Reed-Solomon error correction scheme because it is applied to a polynomial in $Z_2[x]$. If the scheme is applied to a more general polynomial, the process can be even more involved. We illustrate this next.

Example 5.3 Consider the primitive polynomial $p(x) = x^4 + x^3 + 1$ in $Z_2[x]$. For the element $a = x$ in the field $F = Z_2[x]/(p(x))$ of order 16, we list the field elements that correspond to the first 15 powers of a in the following table.

Power	Field Element	Power	Field Element
a^1	a	a^9	$a^2 + 1$
a^2	a^2	a^{10}	$a^3 + a$
a^3	a^3	a^{11}	$a^3 + a^2 + 1$
a^4	$a^3 + 1$	a^{12}	$a + 1$
a^5	$a^3 + a + 1$	a^{13}	$a^2 + a$
a^6	$a^3 + a^2 + a + 1$	a^{14}	$a^3 + a^2$
a^7	$a^2 + a + 1$	a^{15}	1
a^8	$a^3 + a^2 + a$		

Let C be the $RS(15,3)$ code that results from $p(x)$. The following is the generator polynomial $g(x)$ for C.

$$\begin{aligned}
g(x) &= (x-a)(x-a^2)(x-a^3)(x-a^4)(x-a^5)(x-a^6) \\
&= \cdots \\
&= x^6 + a^{12}x^5 + x^4 + a^2x^3 + a^7x^2 + a^{11}x + a^6
\end{aligned}$$

To construct one of the codewords in C, consider the polynomial

$$b(x) = a^8x^8 + a^3x^7 + a^7x^6 + a^9x^5 + x^4 + a^3x^3 + a^6x^2 + a^3x + a^7$$

in $F[x]$. Then

$$\begin{aligned}
c(x) &= b(x)g(x) \\
&= \cdots \\
&= a^8x^{14} + a^{12}x^{13} + a^3x^{12} + a^8x^{11} + a^{12}x^{10} + a^3x^9 + a^8x^8 \\
&\quad + a^{12}x^7 + a^3x^6 + a^4x^5 + a^4x^4 + a^8x^3 + a^{12}x^2 + a^{11}x + a^{13}
\end{aligned}$$

is one of the codewords in C. Suppose $c(x)$ is transmitted and we receive the following polynomial $r(x)$.

$$\begin{aligned}
r(x) &= a^8x^{14} + a^{12}x^{13} + a^3x^{12} + a^8x^{11} + a^{12}x^{10} + a^3x^9 + a^9x^8 \\
&\quad + a^5x^7 + a^{13}x^6 + a^4x^5 + a^4x^4 + a^8x^3 + a^{12}x^2 + a^{11}x + a^{13}
\end{aligned}$$

Note that $r(x)$ contains errors in the x^8, x^7, and x^6 positions. To correct $r(x)$, since C is 3-error correcting, we begin by computing the first six syndromes of $r(x)$. We list these syndromes below.

$$\begin{aligned}
S_1 &= r(a) &&= \cdots &&= 0 \\
S_2 &= r(a^2) &&= \cdots &&= 0 \\
S_3 &= r(a^3) &&= \cdots &&= a^2 \\
S_4 &= r(a^4) &&= \cdots &&= 1 \\
S_5 &= r(a^5) &&= \cdots &&= 1 \\
S_6 &= r(a^6) &&= \cdots &&= a^{12}
\end{aligned}$$

These syndromes yield the following syndrome polynomial $S(z)$.

$$S(z) = a^2z^2 + z^3 + z^4 + a^{12}z^5$$

Constructing the Euclidean algorithm table for $a(z) = z^6$ and $b(z) = S(z)$ (again with numerous calculations omitted), we obtain the following.

Row	Q	R	V
-1	$-$	z^6	0
0	$-$	$S(z)$	1
1	$a^3z + a^6$	$a^7z^4 + a^2z^3 + a^8z^2$	$a^3z + a^6$
2	$a^5z + a^6$	$a^4z^3 + a^3z^2$	$a^8z^2 + a^3z + a$
3	$a^3z + a$	a^7z^2	$a^{11}z^3 + a^{10}z^2 + a^3z + a^5$

Hence, $R(z) = a^7 z^2$ and $V(z) = a^{11}z^3 + a^{10}z^2 + a^3 z + a^5$. By evaluating $V(z)$ at successive powers of a, we can find that the roots of $V(z)$ are a^7, a^8, and a^9. Thus, the positions in $r(x)$ that contain errors are $x^{-7} = x^8$, $x^{-8} = x^7$, and $x^{-9} = x^6$. To determine the coefficients of these terms in the error polynomial, we first note that $V'(z) = a^{11}z^2 + a^3$. We can then determine the coefficients of the terms in the error polynomial as follows.

$$e_6 = \frac{R(a^9)}{V'(a^9)} = \frac{a^{25}}{a^{29} + a^3} = \frac{a^{10}}{a^2} = a^8$$

$$e_7 = \frac{R(a^8)}{V'(a^8)} = \frac{a^{23}}{a^{27} + a^3} = \frac{a^8}{a^5} = a^3$$

$$e_8 = \frac{R(a^7)}{V'(a^7)} = \frac{a^{21}}{a^{25} + a^3} = \frac{a^6}{a} = a^5$$

Therefore, the error in $r(x)$ can be expressed as the error polynomial $e(x) = a^5 x^8 + a^3 x^7 + a^8 x^6$. It can easily be verified that forming $r(x) + e(x)$ yields the codeword $c(x)$. $\blacksquare$

5.3 Proof of Reed-Solomon Error Correction

In this section we verify the Reed-Solomon error correction scheme summarized and illustrated in Section 5.2. As we mentioned in Section 5.2, this verification is extensive, so the reader may wish to postpone this section until completing the remainder of this chapter, or skip this section altogether.

Let F be a field of order 2^n, and let C be an $RS(2^n - 1, t)$ code in $F[x]$. Suppose $c(x) \in C$ is transmitted and we receive the polynomial $r(x) = c(x) + e(x)$ for some nonzero error polynomial $e(x)$ in $F[x]$ of degree less than $2^n - 1$. We will denote this error polynomial by $e(x) = \sum_{j=0}^{m-1} e_j x^j$ with $m = 2^n - 1$ and coefficients $e_j \in F$. To determine $e(x)$ from $r(x)$, we begin by computing the first $2t$ syndromes of $r(x)$. We denote these syndromes as follows.

$$S_i = r(a^i) = e(a^i) = \sum_{j=0}^{m-1} e_j a^{ij} \quad \text{for} \quad i = 1, \ldots, 2t$$

Next, we use these syndromes to construct the syndrome polynomial

$S(z) = \sum_{i=0}^{2t-1} S_{i+1} z^i$. Note then that

$$S(z) = \sum_{i=0}^{2t-1} \sum_{j=0}^{m-1} e_j a^{(i+1)j} z^i = \sum_{j=0}^{m-1} e_j a^j \sum_{i=0}^{2t-1} a^{ij} z^i.$$

Let M be the set of integers that correspond to the error positions in $r(x)$. That is, let $M = \{j \leq m - 1 \mid e_j \neq 0\}$. Note also that

$$S(z) = \sum_{j \in M} e_j a^j \sum_{i=0}^{2t-1} a^{ij} z^i = \sum_{j \in M} e_j a^j \left(\frac{1 - a^{j(2t)} z^{2t}}{1 - a^j z} \right)$$

$$= \sum_{j \in M} \frac{e_j a^j}{1 - a^j z} - \sum_{j \in M} \frac{e_j a^{j(2t+1)} z^{2t}}{1 - a^j z}.$$

Hence, for the polynomials

$$R(z) = \sum_{j \in M} e_j a^j \prod_{\substack{i \in M \\ i \neq j}} (1 - a^i z),$$

$$U(z) = \sum_{j \in M} e_j a^{j(2t+1)} \prod_{\substack{i \in M \\ i \neq j}} (1 - a^i z), \text{ and}$$

$$V(z) = \prod_{i \in M} (1 - a^i z),$$

it follows that

$$S(z) = \frac{R(z)}{V(z)} + \frac{U(z) z^{2t}}{V(z)},$$

or, equivalently,

$$U(z) z^{2t} + V(z) S(z) = R(z).$$

This last equation is called the *fundamental equation*. In this equation, $V(z)$ is called the *error locator* polynomial, $R(z)$ is called the *error evaluator* polynomial, and $U(z)$ is called the *error coevaluator* polynomial. Note that this error locator polynomial $V(z)$ is not the same as the error locator polynomial from the BCH error correction scheme. Note also that

$$(U(z), V(z)) = (R(z), V(z)) = 1.$$

We now consider how to determine the error locator, error evaluator, and error coevaluator polynomials. As we will show, these polynomials are the entries in the Euclidean algorithm table for $a(z) = z^{2t}$ and $b(z) = S(z)$ in the first row j for which $\deg(r_j) < t$. The following results verify this. For convenience, in these results we suppress the variable z whenever possible.

Theorem 5.2 *Suppose $VS + Uz^{2t} = R$ for some syndrome polynomial S, and let V_0, U_0, and R_0 be polynomials that satisfy*

$$V_0 S + U_0 z^{2t} = R_0; \quad \deg(V_0) \le t, \ \deg(U_0) < t, \ \deg(R_0) < t.$$

Then there exists a polynomial $h \in F[z]$ such that $V_0 = hV$, $U_0 = hU$, and $R_0 = hR$. If it is also true that $(V_0, U_0) = 1$, then h is constant.

Proof. Note first that since $VS + Uz^{2t} = R$ and $V_0 S + U_0 z^{2t} = R_0$, it follows that

$$V_0 V S + V_0 U z^{2t} = V_0 R$$

and

$$V V_0 S + V U_0 z^{2t} = V R_0.$$

Hence, by subtraction,

$$(V_0 U - V U_0) z^{2t} = V_0 R - V R_0.$$

By a degree argument we can see that both sides of the preceding equation must be equal to 0. Thus,

$$V_0 U - V U_0 = V_0 R - V R_0 = 0.$$

Since $(V, U) = 1$, then there must exist polynomials $\alpha, \beta \in F[z]$ for which $\alpha V + \beta U = 1$. Therefore,

$$V_0 \alpha V + V_0 \beta U = V_0.$$

But since $V_0 U = V U_0$, then

$$V_0 \alpha V + V \beta U_0 = V_0,$$

or, equivalently,

$$(V_0 \alpha + U_0 \beta) V = V_0.$$

Now, let

$$h = V_0 \alpha + U_0 \beta.$$

Then $hV = V_0$. Also, $hVU = V_0 U = V U_0$ implies $hU = U_0$, and $hVR = V_0 R = V R_0$ implies $hR = R_0$. Finally, since h must divide V_0 and U_0, if $(V_0, U_0) = 1$, then h must be constant. $\blacksquare$

Theorem 5.3 *Suppose $a = z^{2t}$ and $b = S$ for some syndrome polynomial S. In the Euclidean algorithm table for a and b, let j be the first row for which $\deg(r_j) < t$. Define $R_0 = r_j$, $U_0 = u_j$, and $V_0 = v_j$. Then R_0, U_0, and V_0 satisfy all of the conditions in Theorem 5.2.*

Proof. From Equation (1.5) we know that $r_j = u_j z^{2t} + v_j S$. Hence, $R_0 = U_0 z^{2t} + V_0 S$. Furthermore, since $R_0 = r_j$ and $\deg(r_j) < t$, we know that $\deg(R_0) < t$. Now, because $\deg(v_{j-1}) < \deg(v_j) = \deg(V_0)$ and $\deg(r_{j-1}) > \deg(r_j) = \deg(R_0)$, it follows that $\deg(v_{j-1}R_0) < \deg(V_0 r_{j-1})$. But by Equation (1.7), we know that $R_0 v_{j-1} - r_{j-1} V_0 = a = z^{2t}$. Thus, $\deg(r_{j-1}V_0) \le 2t$, and since $\deg(r_{j-1}) \ge t$, it follows that $\deg(V_0) \le t$. Also, since $\deg(u_{j-1}) < \deg(u_j) = \deg(U_0)$, it must be the case that $\deg(R_0 u_{j-1}) < \deg(r_{j-1}U_0)$. But by Equation (1.6) we know that $R_0 u_{j-1} - r_{j-1} U_0 = b = S$. Therefore, $\deg(U_0 r_{j-1}) < 2t$, and since $\deg(r_{j-1}) \ge t$, it follows that $\deg(U_0) < t$. It remains to be shown only that $(V_0, U_0) = 1$. But by Equation (1.8) we know that $u_{j-1}v_j - u_j v_{j-1} = 1$. Hence, $u_{j-1}V_0 - U_0 v_{j-1} = 1$, and $(V_0, U_0) = 1$. ∎

In summary, for a syndrome polynomial $S(z)$, to determine the error locator, error evaluator, and error coevaluator polynomials $V(z)$, $R(z)$, and $U(z)$, we construct the Euclidean algorithm table for $a(z) = z^{2t}$ and $b(z) = S(z)$. At the first row j for which $\deg(r_j) < t$, then $r_j = R_0 = hR(z)$, $u_j = U_0 = hU(z)$, and $v_j = V_0 = hV(z)$. But since $(V_0, U_0) = 1$, then $h = 1$. Hence, $r_j = R(z)$, $u_j = U(z)$, and $v_j = V(z)$. By finding the roots of $V(z)$, we can determine the locations of the errors in the received polynomial as described previously. To find the coefficients of the error polynomial terms, note that since $V(z) = \prod_{i \in M}(1 - a^i z)$, then

$$V'(z) = \sum_{j \in M} -a^j \prod_{\substack{i \in M \\ i \ne j}}(1 - a^i z).$$

And recall we already know that

$$R(z) = \sum_{j \in M} e_j a^j \prod_{\substack{i \in M \\ i \ne j}}(1 - a^i z).$$

By evaluating the preceding polynomials at a^{-j}, we obtain the following.

$$V'(a^{-j}) = -a^j \prod_{\substack{i \in M \\ i \ne j}} \left(1 - a^{(i-j)}\right)$$

$$R(a^{-j}) = e_j a^j \prod_{\substack{i \in M \\ i \ne j}} \left(1 - a^{(i-j)}\right)$$

Hence, $\dfrac{R(a^{-j})}{V'(a^{-j})} = e_j$ reveals the coefficient of x^j in the error polynomial.

5.4   Binary Reed-Solomon Codes

We still have two topics to address regarding Reed-Solomon codes. We stated in the introduction to this chapter that Reed-Solomon codes are uniquely ideal for correcting error bursts, but we have not yet discussed how or why. Also, we stated in Section 5.1 that Reed-Solomon codewords are transmitted as binary vectors, but we have not yet mentioned the actual form in which Reed-Solomon codewords are transmitted. It turns out that these two topics are intimately connected. We will first discuss the form in which Reed-Solomon codewords are transmitted.

Consider the codeword $c(x)$ in the Reed-Solomon code C in Example 5.3. To convert $c(x)$ to a binary vector, we would begin by listing the terms in $c(x)$ as follows with increasing powers of x.

$$\begin{aligned}
c(x) \;=\; & a^{13} + a^{11}x + a^{12}x^2 + a^8x^3 + a^4x^4 + a^4x^5 + a^3x^6 + a^{12}x^7 \\
& + a^8x^8 + a^3x^9 + a^{12}x^{10} + a^8x^{11} + a^3x^{12} + a^{12}x^{13} + a^8x^{14}
\end{aligned}$$

Next, we would write the coefficients in $c(x)$ as follows, using the table in Example 5.3 to express each coefficient as a polynomial in a of degree less than the degree of $p(x)$ with increasing powers of a.

$$\begin{aligned}
c(x) \;=\; & (a + a^2) + (1 + a^2 + a^3)x + (1 + a)x^2 + (a + a^2 + a^3)x^3 \\
& + (1 + a^3)x^4 + (1 + a^3)x^5 + (a^3)x^6 + (1 + a)x^7 \\
& + (a + a^2 + a^3)x^8 + (a^3)x^9 + (1 + a)x^{10} + (a + a^2 + a^3)x^{11} \\
& + (a^3)x^{12} + (1 + a)x^{13} + (a + a^2 + a^3)x^{14}
\end{aligned}$$

Finally, we would express each of these coefficients of $c(x)$ as binary vectors of length four positions by listing in order the binary coefficients of a. For example, we would express the first coefficient $0 + 1a + 1a^2 + 0a^3$ of $c(x)$ as the vector (0110). Using this method, we would convert the entire codeword $c(x)$ into a binary vector of length 60 positions by listing together these binary vectors of length four positions, including four zeros for all terms that could be present in $c(x)$ but have a coefficient of 0 (of which there are none in this codeword). That is, we would convert $c(x)$ to the following binary vector of length 60 positions.

(011010111100011110011001000111000111000111000111000111000111)

It is clear by the fact that Reed-Solomon codewords are converted to binary vectors in this manner why Reed-Solomon codes are ideal for correcting error bursts. Specifically, four errors in the binary equivalent of $c(x)$ constructed above could represent only a single error in $c(x)$. Hence, although $c(x)$ is a codeword in a code that is only 3-error correcting, it may be possible to correct a received vector to $c(x)$ even if 12 errors occur during transmission of the binary equivalent of $c(x)$. More generally, regarding the code C, we would say that provided only one error burst occurs during transmission of the binary equivalent of a codeword in C, we are guaranteed to be able to correct the received vector as long as the error burst is not longer than nine positions. This is because any error burst of length not longer than nine positions in the binary equivalent of a codeword in C could not span more than three of the coefficients in the codeword, while an error burst of length ten positions in the binary equivalent of a codeword in C could span four of the coefficients in the codeword. This statement can be generalized in an obvious manner to apply to any $RS(2^n - 1, t)$ code (see Written Exercise 7).

As we mentioned in the introduction to this chapter, there are many situations in which transmission errors in binary vectors occur naturally in bursts. It is for these situations that Reed-Solomon codes are ideal.

5.5   Reed-Solomon Codes with Maple

In this section we show how Maple can be used to construct codewords and correct errors in the $RS(31, 4)$ code C that results from the primitive polynomial $p(x) = x^5 + x^3 + 1 \in Z_2[x]$.

We begin by including the Maple **linalg** package, entering $p(x)$, and assigning the number $2^5 = 32$ of elements in the field $F = Z_2[x]/(p(x))$ as the variable **fs**.

```
> with(linalg):
> p := x -> x^5 + x^3 + 1:
> Primitive(p(x)) mod 2;
```

$$true$$

```
> fs := 2^(degree(p(x)));
```

$$fs := 32$$

Next, as we did for the BCH code in Section 4.3, we generate and store the field elements in the vector **field** by entering the following commands.

```
> field := vector(fs);
```

$$field := \operatorname{array}(\,1..32, [\,]\,)$$

```
> for i from 1 to fs-1 do
>       field[i] := Powmod(a, i, p(a), a) mod 2:
> od:
> field[fs] := 0:
> evalm(field);
```

$$[a, a^2, a^3, a^4, a^3 + 1, a^4 + a, a^3 + a^2 + 1, a^4 + a^3 + a,$$
$$a^4 + a^3 + a^2 + 1, a^4 + a + 1, a^3 + a^2 + a + 1, a^4 + a^3 + a^2 + a,$$
$$a^4 + a^2 + 1, a + 1, a^2 + a, a^3 + a^2, a^4 + a^3, a^4 + a^3 + 1,$$
$$a^4 + a^3 + a + 1, a^4 + a^3 + a^2 + a + 1, a^4 + a^2 + a + 1,$$
$$a^2 + a + 1, a^3 + a^2 + a, a^4 + a^3 + a^2, a^4 + 1, a^3 + a + 1,$$
$$a^4 + a^2 + a, a^2 + 1, a^3 + a, a^4 + a^2, 1, 0]$$

As we also did for the BCH code in Section 4.3, we establish an association between the polynomial field elements and corresponding powers of a in the table ftable by entering the following commands.

```
> ftable := table():
> for i from 1 to fs-1 do
>       ftable[ field[i] ] := a^i:
> od:
> ftable[ field[fs] ] := 0:
```

5.5.1 Construction of the Codewords

Before constructing any of the codewords in C, we must construct the generator polynomial for C. To do this, we first assign the number $t = 4$ of errors the code is to be able to correct. We can then use the Maple **product** command as follows to construct the generator polynomial for C.

```
> t := 4:
> g := product(x-a^j, j=1..2*t);
```

$$g := (x - a)(x - a^2)(x - a^3)(x - a^4)(x - a^5)(x - a^6)$$
$$(x - a^7)(x - a^8)$$

The process of expanding and simplifying this polynomial so that its coefficients are written in a desirable form (as powers of a) is nontrivial as it requires several conversions between the polynomial field elements and powers of a. Hence, for performing this expansion and simplification, we have provided the user-written procedure **rscoeff**, for which code is given in Appendix C.1. If this procedure is saved as the text file *rscoeff* in the directory from which we are running Maple, then we can include the **rscoeff** procedure in this Maple session by entering the following command.

```
> read rscoeff;
```

We can then see the expanded and simplified form of the generator polynomial $g(x)$ by entering the following command.

```
> g := rscoeff(g, x, p(a), a);
```

$$g := x^8 + a^6 x^7 + a^{27} x^6 + a^4 x^5 + a^{17} x^4 + a^{13} x^3 + a^{14} x^2 + a^2 x + a^5$$

The first two parameters in the preceding command are the polynomial we wish to simplify and the variable used in this polynomial. The final two parameters are the primitive polynomial $p(x)$ in terms of the field element $a = x$ followed by the field element a.

Recall that C is the set of all multiples $b(x)g(x)$ of degree less than 31 with $b(x) \in F[x]$. For example, consider the following polynomial $b(x)$.

```
> b := a^18*x^8 + a^20*x^7 + a^19*x^6 + a^23*x^5 + a^6*x^4
> + a^2*x^3 + a^23*x^2 + a^4*x + a^15:
```

We can construct the codeword $c(x) = b(x)g(x) \in C$ that results from $b(x)$ by entering the following command. Note that we use **rscoeff** so that $c(x)$ will be displayed in a simplified form.

```
> c := rscoeff(b*g, x, p(a), a);
```

$$c := a^{18} x^{16} + a^{14} x^{15} + a^{10} x^{14} + a^{30} x^{13} + a^6 x^{12} + a^{24} x^{11} + a^5 x^{10} + a^{26} x^9 + a^{27} x^8 + a^{30} x^7 + a^{24} x^6 + a^{27} x^5 + a^{21} x^4 + a^9 x^2 + a^{28} x + a^{20}$$

Recall that to transmit this codeword we would convert $c(x)$ into a binary vector using the process described in Section 5.4. To perform this conversion we have provided the user-written procedure **binmess**,[1] for which code is also given in Appendix C.1. If this procedure is saved as the text file *binmess* in the directory from which we are running Maple, we can include the **binmess** procedure in this Maple session as follows.

```
> read binmess;
```

We can then find the binary vector that corresponds to $c(x)$ by entering the following command.

```
> cbin := binmess(c, degree(p(x)), p(a), a, fs-2);
```

$$cbin := [1, 1, 1, 1, 1, 1, 0, 1, 0, 0, 1, 0, 1, 1, 1, 0, 0, 0, 0, 0, 1, 1, 1,$$
$$0, 1, 0, 1, 1, 0, 1, 0, 0, 1, 1, 1, 0, 0, 1, 0, 1, 0, 1, 1, 0, 1, 1, 1, 0,$$

---

[1] The **binmess** procedure uses the Maple **stackmatrix** command. See footnote p. 51 regarding the **stackmatrix** command.

1, 0, 1, 0, 0, 1, 0, 0, 0, 1, 1, 1, 0, 1, 0, 0, 1, 0, 0, 1, 0, 1, 1, 1, 0,
0, 1, 1, 1, 0, 0, 0, 1, 0, 0, 1, 1, 0, 0, 0, 0, 0, 0, 0, 0, 0, 0, 0, 0, 0,
0, 0, 0, 0, 0, 0, 0, 0, 0, 0, 0, 0, 0, 0, 0, 0, 0, 0, 0, 0, 0, 0, 0, 0, 0,
0, 0, 0, 0, 0, 0, 0, 0, 0, 0, 0, 0, 0, 0, 0, 0, 0, 0, 0, 0, 0, 0, 0, 0, 0,
0, 0, 0, 0, 0, 0, 0]

The first parameter in the preceding command is the polynomial for which we wish to find the binary equivalent. The last parameter is the largest possible degree **fs-2** of the codewords in C. Although the codewords in C can be of degree up to 30, the degree of $c(x)$ is only 16. Note that **binmess** recognizes that the terms in $c(x)$ of degrees 17 through 30 have coefficients of zero and inserts appropriate zeros in the resulting binary vector.

5.5.2 Error Correction

Suppose a codeword in C is transmitted as a binary vector and we receive the following vector **rbin**.

```
> rbin := vector([1, 1, 1, 1, 1, 1, 0, 1, 0, 0, 1, 0, 1, 1, 1,
> 0, 0, 0, 0, 0, 1, 1, 1, 0, 1, 0, 1, 1, 0, 1, 0, 0, 1, 1, 1, 0,
> 0, 1, 0, 1, 0, 1, 1, 0, 1, 1, 1, 0, 1, 1, 0, 0, 1, 1, 0, 0, 1,
> 0, 1, 0, 1, 1, 0, 1, 0, 1, 1, 1, 1, 1, 0, 0, 0, 0, 1, 1, 0, 1,
> 1, 0, 1, 1, 0, 0, 1, 0, 0, 1, 1, 0, 0, 0, 0, 1, 1, 1, 0, 0, 1,
> 1, 1, 1, 1, 0, 1, 0, 0, 0, 1, 1, 1, 0, 0, 1, 1, 0, 0, 0, 0, 0,
> 0, 0, 0, 0, 0, 0, 0, 0, 0, 0, 0, 0, 0, 0, 0, 0, 0, 0, 0, 0, 0,
> 0, 0, 0, 0, 0, 0, 0, 0, 0, 0, 0, 0, 0, 0]):
```

To correct this received vector we must first convert the vector to its polynomial equivalent. To help with this conversion we have provided the user-written procedure **bincoeff**, for which code is given in Appendix C.1. Assuming this procedure is saved as the text file *bincoeff* in the directory from which we are running Maple, we can include this procedure in our Maple session as follows.

```
> read bincoeff;
```

Then, by entering the following command we obtain an ordered list of the coefficients in the polynomial equivalent of **rbin**. The first parameter in the following command is the number of binary digits in **rbin** that should be used to form each of these coefficients.

```
> pcoeff := bincoeff(5, rbin);
```

$$pcoeff := [a^4 + a^3 + a^2 + a + 1, a^2 + 1, a^4 + a^3 + a^2 + 1, 0,$$
$$a^4 + a^2 + a + 1, a^4 + a^2 + a, a^4 + a^3 + a^2, a^4 + a^2,$$

$$a^4 + a^2 + a, \; a^4 + a^3 + a + 1, \; a^3 + a^2, \; a^3 + a, \; a^3 + a + 1,$$
$$a^4 + a^3 + a^2 + a + 1, \; a^4, \; a^3 + a^2 + 1, \; a^4 + a + 1, \; a^3 + a^2,$$
$$a^4 + a^3, \; a^4 + a^3 + 1, \; a^4 + a^2 + a + 1, \; a^4 + a^3, \; a^4 + a^3 + 1,$$
$$0, 0, 0, 0, 0, 0, 0, 0]$$

We can then construct a polynomial $r(x)$ with these coefficients by entering the following command.

```
> r := sum('pcoeff[i]*x^(i-1)', 'i'=1..vectdim(pcoeff));
```

$$r := 1 + a + a^3 + a^2 + a^4 + (a^2 + 1) x + (a^4 + a^3 + a^2 + 1) x^2$$
$$+ (a^4 + a^2 + a + 1) x^4 + (a^4 + a^2 + a) x^5 + (a^4 + a^3 + a^2) x^6$$
$$+ (a^4 + a^2) x^7 + (a^4 + a^2 + a) x^8 + (a^4 + a^3 + a + 1) x^9$$
$$+ (a^3 + a^2) x^{10} + (a^3 + a) x^{11} + (a^3 + a + 1) x^{12}$$
$$+ (a^4 + a^3 + a^2 + a + 1) x^{13} + a^4 x^{14} + (a^3 + a^2 + 1) x^{15}$$
$$+ (a^4 + a + 1) x^{16} + (a^3 + a^2) x^{17} + (a^4 + a^3) x^{18}$$
$$+ (a^4 + a^3 + 1) x^{19} + (a^4 + a^2 + a + 1) x^{20} + (a^4 + a^3) x^{21}$$
$$+ (a^4 + a^3 + 1) x^{22}$$

We can use **rscoeff** as follows to simplify the coefficients in the preceding polynomial $r(x)$.

```
> r := rscoeff(r, x, p(a), a);
```

$$r := a^{18} x^{22} + a^{17} x^{21} + a^{21} x^{20} + a^{18} x^{19} + a^{17} x^{18} + a^{16} x^{17}$$
$$+ a^{10} x^{16} + a^7 x^{15} + a^4 x^{14} + a^{20} x^{13} + a^{26} x^{12} + a^{29} x^{11}$$
$$+ a^{16} x^{10} + a^{19} x^9 + a^{27} x^8 + a^{30} x^7 + a^{24} x^6 + a^{27} x^5$$
$$+ a^{21} x^4 + a^9 x^2 + a^{28} x + a^{20}$$

Finally, we enter the following **unapply** command so that we can evaluate $r(x)$ in the usual manner.

```
> r := unapply(r, x);
```

$$r := x \rightarrow a^{18} x^{22} + a^{17} x^{21} + a^{21} x^{20} + a^{18} x^{19} + a^{17} x^{18} + a^{16} x^{17}$$
$$+ a^{10} x^{16} + a^7 x^{15} + a^4 x^{14} + a^{20} x^{13} + a^{26} x^{12} + a^{29} x^{11}$$
$$+ a^{16} x^{10} + a^{19} x^9 + a^{27} x^8 + a^{30} x^7 + a^{24} x^6 + a^{27} x^5$$
$$+ a^{21} x^4 + a^9 x^2 + a^{28} x + a^{20}$$

We will now use the Reed-Solomon error correction scheme to correct $r(x)$. Recall that to correct $r(x)$ we begin by computing the first $2t$ syndromes of $r(x)$. Before doing this we create a vector Sa of length $2t$ positions in which to store the syndromes. In the subsequent loop we generate and store the first $2t$ syndromes of $r(x)$ in Sa. Note that we use the Maple **Rem** function to simplify the syndromes, and that we use **ftable** to find the representations of the syndromes as powers of a. The **if** statement that appears in the loop reduces the syndromes to 1 if they are expressed as a raised to the order of F^*.

```
> Sa := vector(2*t);
```

$$Sa := \text{array}(1..8, [\,])$$

```
> for i from 1 to 2*t do
>       Sa[i] := ftable[ Rem(r(a^i), p(a), a) mod 2 ]:
>       if degree(Sa[i], a) = (fs-1) then
>            Sa[i] := Sa[i]/a^(fs-1):
>       fi:
> od:
> evalm(Sa);
```

$$[\,a^{13}, a^{20}, a^{8}, a^{2}, a^{21}, a^{14}, a^{3}, a^{3}\,]$$

Next, we use the Maple **sum** command to form the resulting syndrome polynomial $S(z)$, and use **unapply** to convert $S(z)$ into a function that we can evaluate in the usual manner.

```
> S := sum('Sa[j+1]*z^j', 'j'=0..2*t-1);
```

$$S := a^{13} + a^{20}\, z + a^{8}\, z^{2} + a^{2}\, z^{3} + a^{21}\, z^{4} + a^{14}\, z^{5} + a^{3}\, z^{6} + a^{3}\, z^{7}$$

```
> S := unapply(S, z);
```

$$S := z \rightarrow a^{13} + a^{20}\, z + a^{8}\, z^{2} + a^{2}\, z^{3} + a^{21}\, z^{4} + a^{14}\, z^{5} + a^{3}\, z^{6} + a^{3}\, z^{7}$$

We must now construct the Euclidean algorithm table for $S(z)$ and the following polynomial $f(z) = z^{2t}$.

```
> f := z^(2*t);
```

$$f := z^{8}$$

To perform the calculations necessary in constructing this table, we have provided the user-written procedure **rseuclid**, for which code is given in Appendix C.1. Assuming this procedure is saved as the text file *rseuclid* in the directory from which we are running Maple, we can include this procedure in our Maple session as follows. (Note: Because **rseuclid** calls and uses the **rscoeff** procedure we discussed previously, the **rscoeff** procedure must be saved as the text file *rscoeff* in the same directory as *rseuclid*.)

```
> read rseuclid;
```

Then the following command causes Maple to construct and display the entries in each row of the Euclidean algorithm table for $S(z)$ and $f(z)$, stopping at the appropriate row for the Reed-Solomon error correction scheme.

The first parameter in this command is the number of errors that can be corrected in the code. The next two parameters are the two polynomials for which we are constructing the table. The fourth parameter is the variable z used in the previous two parameters.

```
> res := rseuclid(t, f, S(z), z, p(a), a);
```

$Q \; = \; , \; a^{28} z + a^{28}, \quad R \; = \; ,$
$\quad a^{23} z^6 + a^{27} z^5 + a^7 z^4 + a^9 z^3 + a^{25} z^2 + a^{26} z + a^{10}, \quad V \; = \; ,$
$\quad a^{28} z + a^{28}, \quad U \; = \; , 1$

$Q \; = \; , \; a^{11} z + a^5, \quad R \; = \; ,$
$\quad a^{17} z^5 + a^{27} z^4 + a^{23} z^3 + a^{24} z^2 + a^5 z + a^{10}, \quad V \; = \; ,$
$\quad a^8 z^2 + a^{12} z + a^{28}, \quad U \; = \; , a^{11} z + a^5$

$Q \; = \; , \; a^6 z + a^{20}, \quad R \; = \; ,$
$\quad a^{26} z^4 + a^{21} z^3 + a^{26} z^2 + a^{27} z + a^{22}, \quad V \; = \; ,$
$\quad a^{14} z^3 + a^{24} z^2 + a^{11} z + a^9, \quad U \; = \; , a^{17} z^2 + a^{23} z + a^4$

$Q \; = \; , \; a^{22} z + a^8, \quad R \; = \; , a^{24} z^3 + a^7 z^2 + a^{16} z + a^{22}, \quad V \; = \; ,$
$\quad a^5 z^4 + z^3 + a^{24} z^2 + a^2 z + a^9, \quad U \; = \; ,$
$\quad a^8 z^3 + a^6 z^2 + a^{28} z + a^{21}$

$res := a^{22} z + a^8, a^{24} z^3 + a^7 z^2 + a^{16} z + a^{22},$
$\quad a^5 z^4 + z^3 + a^{24} z^2 + a^2 z + a^9, a^8 z^3 + a^6 z^2 + a^{28} z + a^{21}$

Note that the preceding process stops at the first row for which the degree of the entry in the **R** column is less than $t = 4$. Note also that the process leaves the vector res containing the entries in the last computed row of the table. Hence, the entries in the table of which we are interested, the polynomials $R(z)$ in the **R** column and $V(z)$ in the **V** column, are the second and third entries in res. We define these entries next as the variables R and V, and use **unapply** to convert each to a function that we can evaluate in the usual manner.

```
> R := res[2];
```

$$R := a^{24} z^3 + a^7 z^2 + a^{16} z + a^{22}$$

```
> R := unapply(R, z);
```

$$R := z \to a^{24} z^3 + a^7 z^2 + a^{16} z + a^{22}$$

```
> V := res[3];
```

$$V := a^5 z^4 + z^3 + a^{24} z^2 + a^2 z + a^9$$

```
> V := unapply(V, z);
```

$$V := z \rightarrow a^5 z^4 + z^3 + a^{24} z^2 + a^2 z + a^9$$

Next, we find the roots of V by trial and error as follows. These commands also show the corresponding error positions in $r(x)$.

```
> for i from 1 to fs-1 do
>         if (Rem(V(a^i), p(a), a) mod 2) = 0 then
>             print(a^i, '    is a root of    ', V(z), '    error
>             position is    ', degree(a^(fs-1)/a^i, a));
>         fi:
> od;
```

a^{15},   *is a root of*   , $a^5 z^4 + z^3 + a^{24} z^2 + a^2 z + a^9$,
     error position is   , 16

a^{16},   *is a root of*   , $a^5 z^4 + z^3 + a^{24} z^2 + a^2 z + a^9$,
     error position is   , 15

a^{17},   *is a root of*   , $a^5 z^4 + z^3 + a^{24} z^2 + a^2 z + a^9$,
     error position is   , 14

a^{18},   *is a root of*   , $a^5 z^4 + z^3 + a^{24} z^2 + a^2 z + a^9$,
     error position is   , 13

To find the coefficients of the terms in the error polynomial, we need to find the derivative of V. We find this derivative next using the Maple **diff** command.

```
> Vp := diff(V(z), z) mod 2;
```

$$Vp := z^2 + a^2$$

```
> Vp := unapply(Vp, z);
```

$$Vp := z \rightarrow z^2 + a^2$$

We can then define the coefficients of the four terms in the error polynomial as follows.

```
> e13 := ftable[ Rem(R(a^18), p(a), a) mod 2 ] /
>           ftable[ Rem(Vp(a^18), p(a), a) mod 2 ]:
> e14 := ftable[ Rem(R(a^17), p(a), a) mod 2 ] /
>           ftable[ Rem(Vp(a^17), p(a), a) mod 2 ]:
> e15 := ftable[ Rem(R(a^16), p(a), a) mod 2 ] /
```

```
>         ftable[ Rem(Vp(a^16), p(a), a) mod 2 ]:
> e16 := ftable[ Rem(R(a^15), p(a), a) mod 2 ] /
>         ftable[ Rem(Vp(a^15), p(a), a) mod 2 ]:
```

Next, we form the error polynomial $e(x)$ that corresponds to $r(x)$.

```
> e := e16*x^16 + e15*x^15 + e14*x^14 + e13*x^13;
```

$$e := a^{14} x^{16} + a x^{15} + a^{12} x^{14} + a^2 x^{13}$$

```
> e := unapply(rscoeff(e, x, p(a), a), x);
```

$$e := x \rightarrow a^{14} x^{16} + a x^{15} + a^{12} x^{14} + a^2 x^{13}$$

Finally, by adding this error polynomial to $r(x)$, we obtain the following corrected codeword $c(x) \in C$.

```
> c := rscoeff(r(x)+e(x), x, p(a), a);
```

$$
\begin{aligned}
c := {} & a^{18} x^{22} + a^{17} x^{21} + a^{21} x^{20} + a^{18} x^{19} + a^{17} x^{18} + a^{16} x^{17} \\
& + a^4 x^{16} + a^{11} x^{15} + a^{23} x^{14} + a^{19} x^{13} + a^{26} x^{12} + a^{29} x^{11} \\
& + a^{16} x^{10} + a^{19} x^9 + a^{27} x^8 + a^{30} x^7 + a^{24} x^6 + a^{27} x^5 \\
& + a^{21} x^4 + a^9 x^2 + a^{28} x + a^{20}
\end{aligned}
$$

The following command shows that $c(x)$ is in C by verifying that $c(a^i) = 0$ for $i = 1, \ldots, 2t$.

```
> seq(Rem(subs(x=a^i, c), p(a), a) mod 2, i=1..2*t);
```

$$0, 0, 0, 0, 0, 0, 0, 0$$

To see the positions in **rbin** that contained errors, we can use the **binmess** routine as follows to find the binary representation of $e(x)$.

```
> ebin := binmess(e(x), degree(p(x)), p(a), a, fs-2);
```

$ebin :=$ [0, 0, 0, 0, 0, 0, 0, 0, 0, 0, 0, 0, 0, 0, 0, 0, 0, 0, 0, 0, 0, 0, 0,
0, 0, 0, 0, 0, 0, 0, 0, 0, 0, 0, 0, 0, 0, 0, 0, 0, 0, 0, 0, 0, 0, 0, 0,.0,
0, 0, 0, 0, 0, 0, 0, 0, 0, 0, 0, 0, 0, 0, 0, 0, 0, 0, 0, 1, 0, 0, 0, 1, 1,
1, 1, 0, 1, 0, 0, 0, 1, 1, 0, 0, 0, 0, 0, 0, 0, 0, 0, 0, 0, 0, 0, 0, 0, 0,
0, 0, 0, 0, 0, 0, 0, 0, 0, 0, 0, 0, 0, 0, 0, 0, 0, 0, 0, 0, 0, 0, 0, 0, 0,
0, 0, 0, 0, 0, 0, 0, 0, 0, 0, 0, 0, 0, 0, 0, 0, 0, 0, 0, 0, 0, 0, 0, 0, 0,
0, 0, 0, 0, 0, 0, 0]

We can then use the Maple **sum** command as follows to see the number of binary errors in **rbin**.

```
> berrors := sum('ebin[i]', 'i'=1..vectdim(ebin));
```

$$berrors := 8$$

Note that although C is only 4-error correcting, we were able to correct the binary vector `rbin` in C even though it contained 8 errors. This is because the binary errors in `rbin` occurred together (*i.e.*, as an error burst) and resulted in only four errors in the corresponding polynomial $r(x)$.

5.6 Reed-Solomon Codes in Voyager 2

In August and September 1977, NASA launched the Voyager 1 and Voyager 2 satellites from Cape Canaveral, Florida. Upon reaching their first destination goals of Jupiter and Saturn, the Voyager satellites provided NASA with the most detailed analyses and visual images of these planets and their moons that had ever been observed. After encountering Jupiter and Saturn, Voyager 2 continued farther into the outer reaches of our solar system and successfully transmitted to Earth data and visual images from Uranus and Neptune. Without the use of a Reed-Solomon code in transmitting these images, the extreme success achieved by Voyager 2 would have been very unlikely. We briefly describe the image transmission process next.

Images transmitted to Earth from outer space are usually digitized into binary strings and sent over a space channel. Voyager 2 digitized its full-color images into binary strings with 15,360,000 positions. Using an uncompressed spacecraft telecommunication system, these binary digits were transmitted one by one to Earth, where the images were then reconstructed. This uncompressed system was the most reliable system available when Voyager 2 was launched, and was satisfactory for transmitting images to Earth from Jupiter and Saturn. However, when Voyager 2 arrived at Uranus in January 1986, it was about twice as far from Earth as it had been when at Saturn. Since the transmission of binary digits to Earth had already been stretched to a very slow rate from Saturn (around 44,800 digits per second), a new transmission scheme was needed in order for NASA to be able to receive a large number of images from Uranus.

The problem of image transmission from Uranus was solved through the work of Robert Rice at California Institute of Technology's Jet Propulsion Laboratory. Rice developed an algorithm that implemented a compressed spacecraft telecommunication system which reduced by a factor of 2.5 the amount of data needed to transmit a single image from Uranus without causing any loss in image quality. However, there was a problem with Rice's algorithm. During the long transmissions through space, the compressed binary strings experienced errors much more frequently than the uncompressed strings, and Rice's algorithm was very sensitive to binary errors. In fact, if a received binary transmission from Uranus contained even a single error, the resulting image would be completely ruined. After considerable study, it was discovered that the binary errors that occurred during the long transmissions through space usually occurred in bursts. To account for these error bursts, a new system was designed in Voyager 2 for converting images into binary strings that utilized a Reed-Solomon code. The binary strings were compressed and transmitted back to Earth, and then uncompressed using Rice's algorithm and corrected using the Reed-Solomon error correction scheme. This process was highly successful. The specific Reed-Solomon code used in Voyager 2 is mentioned in Maple Exercise 4.

After leaving Uranus, Voyager 2 continued its journey through space. In August 1989 the satellite transmitted data and visual images to Earth that provided NASA with most of the information currently known about Neptune. At present, the Voyager 2 satellite is still in operation and is still providing NASA with invaluable information about our solar system.

In addition to being used in the transmission of images through space, Reed-Solomon codes have a rich assortment of other applications, and are claimed to be the most frequently used digital error-correcting codes in the world. As we mentioned in the introduction to this chapter, Reed-Solomon codes are used extensively in the encoding of various types of information on compact discs. Also, Reed-Solomon codes have played an integral role in the development of high-speed supercomputers. In the future, Reed-Solomon codes will be an important tool for dealing with complex communication and information transfer systems.

Written Exercises

1. Let C be the $RS(7, 2)$ code that results from the primitive polynomial $p(x) = x^3 + x + 1 \in Z_2[x]$.

   (a) Construct and simplify the generator polynomial for C.

   (b) Construct two of the codewords in C.

   (c) Convert the codewords you constructed in part (b) into binary vectors using the process described in Section 5.4.

   (d) Find the maximum error burst length that is guaranteed to be correctable in C.

2. Correct the following received polynomials in the Reed-Solomon code C in Written Exercise 1.

   (a) $r(x) = a^5x^6 + ax^5 + a^6x^4 + ax^2 + a^4x + a^5$

   (b) $r(x) = a^4x^6 + a^3x^5 + x^4 + a^4x^3 + a^2x^2 + a^2x + a^3$

   (c) $r(x) = ax^5 + a^4x^4 + a^6x^3 + a^3x^2 + a^5x + a^4$

3. Use the Reed-Solomon error correction scheme to correct the following received polynomial $r(x)$ in the BCH code C in Example 4.3.

$$r(x) = 1 + x^4 + x^5 + x^7 + x^8 + x^{11} + x^{13}$$

4. Correct the following received polynomials in the Reed-Solomon code C in Example 5.3.

   (a) $r(x) = ax^{12} + a^2x^{11} + a^8x^{10} + a^6x^9 + x^8 + a^7x^7 + a^{10}x^6$
   $+ a^6x^5 + a^4x^4 + a^7x^3 + ax^2 + a^6x$

   (b) $r(x) = a^5x^{14} + a^{14}x^{13} + a^3x^{12} + a^4x^{11} + a^{14}x^{10} + a^4x^9$
   $+ a^{11}x^8 + a^7x^7 + a^{11}x^6 + a^9x^5 + a^{14}x^4 + a^7x^3$
   $+ a^{10}x + a^5$

5. Use the reverse of the process described in Section 5.4 to convert the following binary vector to the polynomial codeword it represents in the Reed-Solomon code C in Example 5.3.

   (1101010100001110001110101011111101011100100111001000100111101)

6. Prove Theorem 5.1.

7. Let C be an $RS(2^n - 1, t)$ code. In terms of n and t, find the maximum error burst length that is guaranteed to be correctable in C.

Maple Exercises

1. Correct the following received binary vector r in the $RS(31, 4)$ code C considered in Section 5.5.

   $$r = (1010010001010011010000010101010011110110000101010010$$
   $$11011111010001010011010000011101111100000001000001$$
   $$0001101100100000000000000000000000000000000000000)$$

2. Let C be the $RS(127, 4)$ code that results from the primitive polynomial $p(x) = x^7 + x + 1 \in Z_2[x]$.

   (a) Construct and simplify the generator polynomial $g(x)$ for C.

   (b) Construct the codeword $b(x)g(x) \in C$ that results from the following polynomial $b(x)$.

   $$b(x) = a^{30}x^{100} + a^2 x^{51} + a^{91}x^2 + a^3$$

   (c) Convert the codeword $b(x)g(x)$ in part (b) into a binary vector using the process described in Section 5.4 (i.e., using the **binmess** procedure).

   (d) Find the maximum error burst length that is guaranteed to be correctable in C.

3. Correct the following received polynomials in the Reed-Solomon code C in Maple Exercise 2.

   (a) $r(x) = a^{100}x^{22} + a^{10}x^{21} + a^{100}x^{20} + a^{28}x^{19} + a^{114}x^{18}$
   $$+ a^{35}x^{17} + a^{81}x^{16} + a^{95}x^{15} + a^{56}x^{14} + a^{59}x^{13}$$
   $$+ a^{38}x^{12} + a^{83}x^{11} + a^{42}x^{10}$$

   (b) $r(x) = a^{10}x^{108} + a^{60}x^{107} + a^{49}x^{106} + a^{115}x^{105} + a^{18}x^{104}$
   $$+ a^{124}x^{103} + a^{67}x^{102} + a^{87}x^{101} + a^{46}x^{100}$$

   (c) $r(x) = a^{100}x^{81} + a^{23}x^{80} + a^{12}x^{79} + a^{78}x^{78} + a^{108}x^{77}$
   $$+ a^{80}x^{76} + a^{25}x^{75} + a^{50}x^{74} + a^9 x^{73}$$

4. The Reed-Solomon code used in the Voyager 2 satellite (see discussion in Section 5.6) was the $RS(255, 16)$ code C that results from the primitive polynomial $p(x) = x^8 + x^4 + x^3 + x^2 + 1 \in Z_2[x]$. Construct several of the codewords in C as polynomials or binary vectors. Also, illustrate the Reed-Solomon error correction scheme in C by correcting several received binary vectors or polynomials that contain errors. Write a summary of your results.

Chapter 6

Algebraic Cryptography

Cryptography is the study of techniques that can be used to disguise a message so that only the intended recipient can remove the disguise and read it. The simplest way to disguise a message is to replace every occurrence of each specific character with a different character. This method for disguising a message yields what we will call a *substitution cipher*. Since substitution ciphers appear as puzzles in many newspapers and puzzle books, they are obviously relatively easy to "break" and should not be used when sending "top secret" information. In the next three chapters, we discuss some disguising techniques that involve applying mathematical operations to messages. Using mathematics to disguise messages gives us the ability to create disguising techniques that are increasingly more difficult to break by simply choosing mathematical operations that are increasingly more complex. Because the disguising techniques we discuss involve applying mathematical operations to messages, these techniques are examples of *algebraic* cryptography.

6.1 Some Elementary Cryptosystems

We will call an undisguised message a *plaintext* and a disguised message a *ciphertext*. Also, we will call the process of converting a plaintext to a ciphertext the *encryption* or *encipherment* of the message, and we will call the reverse process the *decryption* or *decipherment* of the message.

Since we will encipher messages by applying mathematical operations, our plaintext characters will have to have some kind of mathematical structure. We will give our messages the structure of a ring so that we can both add and multiply message characters.

Definition 6.1 *A cryptosystem consists of the following:*

1. *An alphabet L that contains all characters that can be used in messages (letters, numerals, punctuation marks, blank spaces, etc.),*

2. *A commutative ring R with identity such that $|R| = |L|$,*

3. *Bijections $\alpha : L \to R$ and $f : R \to R$.*

The idea we will take is that to encipher a plaintext that is expressed as a list of elements in L, we first use α to convert the plaintext into a list of elements in R. We can then form the ciphertext by applying f to the plaintext ring elements and, if desired, use α^{-1} to convert the ciphertext back to a list of elements in L. We can recover the plaintext from the ciphertext by repeating this procedure using f^{-1} instead of f. So that only the intended recipient of the message can recover the plaintext, only the intended recipient can know f^{-1}. We will always assume that everything else in a cryptosystem, with the obvious exception of f, is public knowledge. (It is true in some cryptosystems that f can be public knowledge without revealing f^{-1}. Such cryptosystems are called *public-key* systems. We discuss two public-key cryptosystems, the well-known RSA and ElGamal systems, in Chapters 7 and 8.)

For simplicity, in this chapter we will assume that all messages are written in the alphabet $L = \{A, B, \dots, Z\}$. Also, we will take $R = Z_{26}$ and let $\alpha : L \to R$ be given by $\alpha(A) = 0, \alpha(B) = 1, \dots, \alpha(Z) = 25$. For reference, we list the correspondence for α below.

A	B	C	D	E	F	G	H	I	J	K	L	M	N	O
0	1	2	3	4	5	6	7	8	9	10	11	12	13	14

P	Q	R	S	T	U	V	W	X	Y	Z
15	16	17	18	19	20	21	22	23	24	25

We now consider some cryptosystems with different types of encryption methods. That is, we consider some cryptosystems with different types of bijections $f : R \to R$.

Encryption Method 1: Choose $f : R \to R$ by $f(x) = ax \bmod |R|$ for some $a \in R$ with $(a, |R|) = 1$.

Example 6.1 Let $f(x) = 3x \bmod 26$. Then the message "ATTACK AT DAWN" enciphers as follows.

		A	T	T	A	C	K	A	T	D	A	W	N
α	$\Rightarrow$	0	19	19	0	2	10	0	19	3	0	22	13
f	$\Rightarrow$	0	5	5	0	6	4	0	5	9	0	14	13
α^{-1}	$\Rightarrow$	A	F	F	A	G	E	A	F	J	A	O	N

Hence, the corresponding ciphertext is "AFFAGEAFJAON". To decipher this message, we repeat the same procedure using the inverse function $f^{-1}(x) = a^{-1}x \bmod |R| = 9x \bmod 26$ instead of f. (Note: We can determine a^{-1} in general by using the Euclidean algorithm as illustrated in Section 7.1.) This message deciphers, of course, as follows.

		A	F	F	A	G	E	A	F	J	A	O	N
α	$\Rightarrow$	0	5	5	0	6	4	0	5	9	0	14	13
f^{-1}	$\Rightarrow$	0	19	19	0	2	10	0	19	3	0	22	13
α^{-1}	$\Rightarrow$	A	T	T	A	C	K	A	T	D	A	W	N

Note that $3^{-1} = 9 \bmod 26$ because $3 \cdot 9 = 27 = 1 \bmod 26$. Note also that we are guaranteed a multiplicative inverse of $a = 3$ exists modulo $|R| = 26$ because of the requirement that $(a, |R|) = 1$ (see Written Exercise 10). ∎

We will say that any person except the intended recipient who tries to decipher an encrypted message is an *intruder* and is attempting to *break* the cryptosystem. Two people wishing to exchange a secret message would certainly want to use a cryptosystem that an intruder would find difficult to break. However, breaking a cryptosystem is not generally as difficult as it may at first appear. Recall that everything in a cryptosystem is assumed to be public knowledge except f and f^{-1}, and in practice it is usually assumed that even the form of f is publicly known and that only the parameters in f are unknown to intruders. We will call the parameters in f the *keys* of the cryptosystem because if an intruder is able to find these parameters, the intruder should then be able to determine f^{-1} (*i.e.*, "unlock" the system).

The cryptosystem in Example 6.1 has $a = 3$ as its only key. As you may have supposed, this system is not very secure. However, this is not because it has only a single key or because it is so easy to use for enciphering messages. Recall we would assume that intruders know everything about this system except the parameters in f and f^{-1}. Indeed, we would even assume that intruders know $f(x) = ax \bmod 26$ for some $a \in Z_{26}$ with $(a, |R|) = 1$. With a function f of this form, an intruder could very quickly determine the key by trial and error. Specifically, because the only elements in Z_{26} that are relatively prime to 26 are $\{1, 3, 5, 7, 9, 11, 15, 17, 19, 21, 23, 25\}$, then one of these elements must be the key. An intruder could very easily take each of these key candidates separately, form the corresponding inverse functions f^{-1}, and use each f^{-1} to decipher the encrypted message. Most likely, only one of these decipherments, the correct plaintext, will make any sense. Even if the calculations are done with a pencil and paper, an intruder could break this system in only a few minutes.

One obvious way to enhance the security of the cryptosystem in Example 6.1 is to include a constant term in the function f. We summarize this in general as our second encryption method.

Encryption Method 2: Choose $f : R \rightarrow R$ by $f(x) = ax + b \bmod |R|$ for some $a, b \in R$ with $(a, |R|) = 1$.

Example 6.2 Let $f(x) = 3x + 4 \bmod 26$. Then the message "ATTACK AT DAWN" enciphers as follows.

	A	T	T	A	C	K	A	T	D	A	W	N
$\alpha \Rightarrow$	0	19	19	0	2	10	0	19	3	0	22	13
$f \Rightarrow$	4	9	9	4	10	8	4	9	13	4	18	17
$\alpha^{-1} \Rightarrow$	E	J	J	E	K	I	E	J	N	E	S	R

Hence, the corresponding ciphertext is "EJJEKIEJNESR". To decipher this message, we repeat the same procedure using the inverse function $f^{-1}(x) = a^{-1}(x - b) \bmod |R| = 9(x - 4) \bmod 26$ instead of f. ∎

The cryptosystem in Example 6.2 has two keys, $a = 3$ and $b = 4$. While this system is more secure than the cryptosystem in Example 6.1, it is still not a very secure system. We would assume that an intruder who intercepts the ciphertext in Example 6.2 would know that $f(x) = ax + b \bmod 26$ for some $a, b \in Z_{26}$ with $(a, |R|) = 1$. Hence, since a must be one of the twelve elements in $\{1, 3, 5, 7, 9, 11, 15, 17, 19, 21, 23, 25\}$, and b must be one of the 26 elements in Z_{26}, there will only be $12 \cdot 26 = 312$ possible pairs (a, b) of keys. An intruder using only a hand-held calculator could easily test each of these pairs of key candidates in only a few hours. More importantly, an intruder using a computer that can perform millions of operations per second could test each of these pairs of key candidates immediately. Thus, although the cryptosystem in Example 6.2 has more key candidates than the system in Example 6.1, it can still be broken very easily.

We have shown that neither of the cryptosystems described in this section are secure by presenting rather simple mathematical methods for breaking them. We can also see that these systems are not secure because they both just yield substitution ciphers. And the systems described in this section are actually easier to break than nonmathematical substitution ciphers, for in each case our general procedure for breaking the system can easily be programmed on a computer, while breaking nonmathematical substitution ciphers generally requires somewhat time-consuming frequency analysis (*i.e.*, trial and error under the assumption that the most frequently occurring ciphertext characters correspond to the most commonly used plaintext characters). However, the cryptosystems described in this section were not presented in an attempt to illustrate secure systems, but rather because they will be generalized in Sections 6.2 and 6.4 into well-known systems that can be designed with any desired level of security.

6.2 The Hill Cryptosystem

The cryptosystems described in Section 6.1 have been known for many years. In fact, a variation of our second encryption method in Section 6.1 was used in ancient Rome by Julius Caesar, who supposedly invented it himself (see Written Exercise 1). While the advent of calculators and computers have rendered these systems obsolete, generalizations of these systems that use matrices as keys instead of scalars can still be constructed with any desired level of security. The generalization of our first encryption method was first described by Lester Hill in 1929. We summarize this in general as our next encryption method.

Hill Encryption Method: Let A be an $n \times n$ invertible matrix over R (*i.e.*, such that $(\det A, |R|) = 1$). Group the plaintext into row vectors P_i of length n, and define $f : R^n \to R^n$ by $f(P_i) = P_i A$ with each entry taken modulo $|R|$. The resulting rows listed together form the ciphertext.

Example 6.3 In this example, we use the Hill encryption method to encipher the message "MEET AT SEVEN". We begin by converting the message into a list of elements in Z_{26}.

	M	E	E	T	A	T	S	E	V	E	N
$\alpha \Rightarrow$	12	4	4	19	0	19	18	4	21	4	13

We will use the following 2×2 key matrix A to encipher this message.

$$A = \begin{bmatrix} 2 & 5 \\ 1 & 4 \end{bmatrix}$$

Note that A is invertible over Z_{26} since $(\det A, 26) = (3, 26) = 1$. To form the ciphertext, we group the plaintext into row vectors P_i of length 2, and compute $P_i A$ for all i with each entry taken modulo 26. For example, the first ciphertext vector is computed as follows.

$$
\begin{aligned}
P_1 A &= (12, 4) \begin{bmatrix} 2 & 5 \\ 1 & 4 \end{bmatrix} \\
&= (28, 76) \\
&= (2, 24)
\end{aligned}
$$

The remaining ciphertext vectors are computed as follows. (Since the message does not completely fill the last plaintext vector P_6, we fill this vector with an arbitrary element from Z_{26}.)

$$
\begin{aligned}
P_2 A &= (4, 19)A &= \cdots &= (1, 18) \\
P_3 A &= (0, 19)A &= \cdots &= (19, 24) \\
P_4 A &= (18, 4)A &= \cdots &= (14, 2) \\
P_5 A &= (21, 4)A &= \cdots &= (20, 17) \\
P_6 A &= (13, 25)A &= \cdots &= (25, 9)
\end{aligned}
$$

Hence, the entire encipherment is

	M	E	E	T	A	T	S	E	V	E	N	
$\alpha \Rightarrow$	12	4	4	19	0	19	18	4	21	4	13	25
$f \Rightarrow$	2	24	1	18	19	24	14	2	20	17	25	9
$\alpha^{-1} \Rightarrow$	C	Y	B	S	T	Y	O	C	U	R	Z	J

and the ciphertext is "CYBSTYOCURZJ". Although the last ciphertext character is in a position beyond the last plaintext character, it must be retained as it is necessary for decipherment. ∎

One thing we can notice immediately from Example 6.3 is that the Hill encryption method does not in general yield a substitution cipher. Also, enciphering messages with the Hill system requires nothing more than some matrix multiplication with an invertible key matrix A over R. A matrix A over R is invertible if and only if the determinant of A has a multiplicative inverse in R (see Written Exercise 9). For $R = Z_k$, this is equivalent to $(\det A, k) = 1$ (see Written Exercise 10).

To decipher a message that has been encrypted using the Hill system with an $n \times n$ key matrix A, we would group the ciphertext into row vectors C_i of length n and compute $f^{-1}(C_i) = C_i A^{-1} \bmod |R|$ for all i. The matrix A^{-1} over R can be determined in general by the well-known formula

$$
A^{-1} = \frac{1}{\det A} (\operatorname{adj} A) \tag{6.1}
$$

where adj A represents the adjoint of A. To determine adj A, we would first need to find the cofactors of A. These cofactors are defined as

$$C_{ij} = (-1)^{i+j} M_{ij} \qquad i, j = 1, \ldots, n$$

where M_{ij} is the determinant of the matrix obtained by deleting the ith row and j^{th} column from A. Using these cofactors, the adjoint of A is defined as follows.

$$\text{adj } A = \begin{bmatrix} C_{11} & C_{21} & \cdots & C_{n1} \\ C_{12} & C_{22} & \cdots & C_{n2} \\ \vdots & \vdots & & \vdots \\ C_{1n} & C_{2n} & \cdots & C_{nn} \end{bmatrix}$$

That is, adj A is the transpose of the matrix of cofactors of A.

Example 6.4 To decipher the message in Example 6.3, we would first need to find the inverse of the key matrix A. Using (6.1), we can determine this inverse as follows.

$$A^{-1} = \frac{1}{3} \begin{bmatrix} 4 & -5 \\ -1 & 2 \end{bmatrix}$$

$$= 9 \begin{bmatrix} 4 & 21 \\ 25 & 2 \end{bmatrix}$$

$$= \begin{bmatrix} 36 & 189 \\ 225 & 18 \end{bmatrix}$$

$$= \begin{bmatrix} 10 & 7 \\ 17 & 18 \end{bmatrix}$$

Then to recover the plaintext, we group the ciphertext into row vectors C_i of length 2 and compute $C_i A^{-1}$ for all i with each entry taken modulo 26. For example, the first plaintext vector is recovered as follows.

$$C_1 A^{-1} = (2, 24) \begin{bmatrix} 10 & 7 \\ 17 & 18 \end{bmatrix}$$

$$= (428, 446)$$

$$= (12, 4)$$

The remaining plaintext vectors are recovered as follows.

$$
\begin{array}{llllll}
C_2 A^{-1} & = & (1, 18) A^{-1} & = & \cdots & = & (4, 19) \\
C_3 A^{-1} & = & (19, 24) A^{-1} & = & \cdots & = & (0, 19) \\
C_4 A^{-1} & = & (14, 2) A^{-1} & = & \cdots & = & (18, 4) \\
C_5 A^{-1} & = & (20, 17) A^{-1} & = & \cdots & = & (21, 4) \\
C_6 A^{-1} & = & (25, 9) A^{-1} & = & \cdots & = & (13, 25)
\end{array}
$$

Applying α^{-1} to the entries in these plaintext vectors will reveal the original message. (Because we chose an arbitrary element from Z_{26} to fill the last plaintext vector P_6 in Example 6.3, there will be an extra character at the end of this message.)   ∎

We now consider how an intruder could break the cryptosystem in Example 6.3. We would assume that an intruder who intercepts the ciphertext in Example 6.3 would know that the ciphertext was formed by grouping the plaintext into row vectors P_i of length 2 and multiplying each P_i by some 2×2 invertible key matrix A over Z_{26}. Although the requirement that A be invertible does not impose any specific restrictions on any of the four individual entries in A, an intruder would at least know that each of these entries must be elements in Z_{26}. Hence, to find A by trial and error, an intruder would have to test a maximum of only $26^4 = 456976$ possible key matrices. While it would not be feasible for an intruder to test all of these possible key matrices by hand or even with a calculator, an intruder using a computer that can perform millions of operations per second could test these possible key matrices very quickly and easily. Hence, the Hill system with a 2×2 key matrix A does not yield a very secure system. However, if A were chosen of size 3×3, then there would be $26^9 = 5429503678976$ possible key matrices for an intruder to test; and if A were chosen of size 5×5 there would be $26^{25} = 2.37 \times 10^{35}$ possible key matrices. Thus, even with a relatively small key matrix, the Hill system yields a reasonable amount of security. And the Hill system can be used to obtain any desired level of security by simply choosing a key matrix that is sufficiently large.

The Hill system does have a vulnerability we should mention. It is not unreasonable to suppose that an intruder who intercepts a ciphertext formed using the Hill system might know or be able to guess a small part of the plaintext. For example, the intruder may know from whom the message originated and correctly guess that the last few characters in the plaintext were the originator's name or signature. It turns out that it may be possible for an intruder to break the Hill system relatively easily if the intruder knows or is able to correctly guess a small part of the plaintext. More specifically, if the Hill system is used with an $n \times n$ key matrix, it may be possible for an intruder to break the system relatively easily if the intruder knows or is able to correctly guess n^2 characters from the plaintext. We illustrate this in the following example.

Example 6.5 Suppose an intruder intercepts the ciphertext in Example 6.3 and somehow knows or guesses that the last four ciphertext letters were produced by the plaintext letters "VENZ". That is, suppose the intruder knows or guesses the following from the encipherment in Example 6.3.

									V	E	N	Z
$\alpha \Rightarrow$									21	4	13	25
$f \Rightarrow$	2	24	1	18	19	24	14	2	20	17	25	9
$\alpha^{-1} \Rightarrow$	C	Y	B	S	T	Y	O	C	U	R	Z	J

Since the intruder would know that the plaintext was enciphered using some 2×2 key matrix

$$A = \begin{bmatrix} a & b \\ c & d \end{bmatrix}$$

over Z_{26}, then the intruder would know that

$$(21,4)\begin{bmatrix} a & b \\ c & d \end{bmatrix} = (20,17) \quad \text{and} \quad (13,25)\begin{bmatrix} a & b \\ c & d \end{bmatrix} = (25,9)$$

with $a, b, c, d \in Z_{26}$. The preceding two matrix equations are equivalent to the following single matrix equation.

$$\begin{bmatrix} 21 & 4 \\ 13 & 25 \end{bmatrix}\begin{bmatrix} a & b \\ c & d \end{bmatrix} = \begin{bmatrix} 20 & 17 \\ 25 & 9 \end{bmatrix} \tag{6.2}$$

If the intruder could determine a unique solution to this equation over Z_{26}, then this solution would necessarily be the key matrix A for the system. Note that since

$$\begin{vmatrix} 21 & 4 \\ 13 & 25 \end{vmatrix} = 473 = 5$$

has an inverse in Z_{26}, then (6.2) has a unique solution over Z_{26}. Using (6.1), the intruder could find this solution as follows.

$$\begin{bmatrix} a & b \\ c & d \end{bmatrix} = \begin{bmatrix} 21 & 4 \\ 13 & 25 \end{bmatrix}^{-1}\begin{bmatrix} 20 & 17 \\ 25 & 9 \end{bmatrix}$$

$$= \frac{1}{5}\begin{bmatrix} 25 & -4 \\ -13 & 21 \end{bmatrix}\begin{bmatrix} 20 & 17 \\ 25 & 9 \end{bmatrix}$$

$$= 21\begin{bmatrix} 25 & 22 \\ 13 & 21 \end{bmatrix}\begin{bmatrix} 20 & 17 \\ 25 & 9 \end{bmatrix}$$

$$= \begin{bmatrix} 22050 & 13083 \\ 16485 & 8610 \end{bmatrix}$$

$$= \begin{bmatrix} 2 & 5 \\ 1 & 4 \end{bmatrix}$$

Note that $5^{-1} = 21 \bmod 26$ because $5 \cdot 21 = 105 = 1 \bmod 26$. Note also that the last result is the key matrix A from Example 6.3. The intruder could then find A^{-1} and decipher the rest of the ciphertext. ∎

For a message enciphered using the Hill system with an $n \times n$ key matrix, one obvious problem an intruder could encounter when trying to break the system in the way illustrated in Example 6.5 is that even if the intruder knows or correctly guesses n^2 characters from the plaintext, the analogue to (6.2) may not have a unique solution. And even if (6.2) has a unique solution, it may not be possible to find this solution in the way illustrated in Example 6.5. For example, if an intruder intercepts the ciphertext in Example 6.3 and correctly guesses that the first four ciphertext letters were produced by the plaintext letters "MEET", the analogue to (6.2) would be the following.

$$\begin{bmatrix} 12 & 4 \\ 4 & 19 \end{bmatrix} \begin{bmatrix} a & b \\ c & d \end{bmatrix} = \begin{bmatrix} 2 & 24 \\ 1 & 18 \end{bmatrix} \tag{6.3}$$

But since

$$\begin{vmatrix} 12 & 4 \\ 4 & 19 \end{vmatrix} = 212 = 4$$

does not have a multiplicative inverse in Z_{26}, then even if (6.3) has a unique solution over Z_{26} it will not be possible to find this solution in the way illustrated in Example 6.5.

6.3 The Hill Cryptosystem with Maple

In this section we show how Maple can be used to encipher and decipher messages using the Hill cryptosystem.

We begin by establishing the correspondence α between the alphabet letters and ring elements. To do this, we construct the following array **letters** containing the alphabet letters.

```
> letters := array(0..25, [A, B, C, D, E, F, G, H, I, J, K, L,
> M, N, O, P, Q, R, S, T, U, V, W, X, Y, Z]):
```

We can then access the alphabet letters from their positions in this array, with the first letter being in position 0. For example, because C is the third letter in this array, the letter C is returned when its corresponding ring element is entered as follows.

```
> letters[2];
```

$$C$$

So that we can also access the ring elements by entering the alphabet letters, we will also establish the correspondence α in a table. We first create the table `ltable`.

```
> ltable := table():
```

We can then use the array `letters` to establish the correspondence α in `ltable` by entering the following commands.

```
> for i from 0 to 25 do
>     ltable[ letters[i] ] := i:
> od:
```

Using `ltable`, we can access the ring elements by entering the corresponding letters. For example, the ring element that corresponds to the letter C is returned by the following command.

```
> ltable[C];
```

$$2$$

We now show how Maple can be used to encipher the message "RENDEZVOUS AT NOON". We will use the following 3×3 key matrix **A**.

```
> with(linalg):
> A := matrix( [ [11,6,8], [0,3,14], [24,0,9] ] );
```

$$A := \begin{bmatrix} 11 & 6 & 8 \\ 0 & 3 & 14 \\ 24 & 0 & 9 \end{bmatrix}$$

By entering the following commands, we verify that **A** is a valid key matrix by verifying that the determinant of **A** is relatively prime to the number of alphabet letters.

```
> d := det(A) mod 26;
```

$$d := 21$$

```
> gcd(d, 26);
```

$$1$$

Next, we enter the plaintext as the vector `ptext`. Note that we include two extra letters in this vector so that the number of plaintext characters will be a multiple of the number of rows of **A**.

```
> ptext := vector([R, E, N, D, E, Z, V, O, U, S, A, T, N, O, O,
> N, A, A]);
```

$$ptext := [R, E, N, D, E, Z, V, O, U, S, A, T, N, O, O, N, A, A]$$

The following command returns the number of plaintext characters.

```
> vectdim(ptext);
```

$$18$$

By entering the following loop, we convert the list **ptext** of plaintext letters into a list of ring elements. We then use **evalm** to display the result.

```
> for i from 1 to vectdim(ptext) do
>     ptext[i] := ltable[ ptext[i] ]:
> od:
> evalm(ptext);
```

$$[\,17,\, 4,\, 13,\, 3,\, 4,\, 25,\, 21,\, 14,\, 20,\, 18,\, 0,\, 19,\, 13,\, 14,\, 14,\, 13,\, 0,\, 0\,]$$

Before enciphering this message, we must group the plaintext ring elements into row vectors with the same number of positions as the number of rows of **A**. To do this, we first assign the number of rows of **A** as **blocksize**.

```
> blocksize := rowdim(A);
```

$$blocksize := 3$$

Next, we assign the number of row vectors into which we will group the plaintext ring elements as **numblocks**.

```
> numblocks := vectdim(ptext)/blocksize;
```

$$numblocks := 6$$

By entering the following **matrix** command, we group the plaintext ring elements from the vector **ptext** into row vectors of length **blocksize** and place these row vectors in order as rows in the matrix **pmatrix**.

```
> pmatrix := matrix(numblocks, blocksize, ptext);
```

$$pmatrix := \begin{bmatrix} 17 & 4 & 13 \\ 3 & 4 & 25 \\ 21 & 14 & 20 \\ 18 & 0 & 19 \\ 13 & 14 & 14 \\ 13 & 0 & 0 \end{bmatrix}$$

Because **pmatrix** contains all of the plaintext row vectors, we can find all of the ciphertext row vectors at once by multiplying **pmatrix** by **A**. By entering the following command, we compute this product and define the result as **cmatrix**. Note that in this command we use the Maple **map** procedure to reduce the entries in `cmatrix` modulo 26. Note also that we use &* for the matrix multiplication, and **evalm** to display the result.

```
> cmatrix := map(m -> m mod 26, evalm(pmatrix &* A));
```

$$cmatrix := \begin{bmatrix} 5 & 10 & 23 \\ 9 & 4 & 19 \\ 9 & 12 & 24 \\ 4 & 4 & 3 \\ 11 & 16 & 10 \\ 13 & 0 & 0 \end{bmatrix}$$

We can use the Maple **convert** command as follows to list the rows in `cmatrix` in order as the vector `ctext`.

```
> ctext := convert(cmatrix, vector);
```

$$ctext := [5, 10, 23, 9, 4, 19, 9, 12, 24, 4, 4, 3, 11, 16, 10, 13, 0, 0]$$

The preceding output shows the ciphertext expressed as a list of ring elements. By entering the following loop, we convert this list of ring elements into a list of alphabet letters.

```
> for i from 1 to vectdim(ctext) do
>       ctext[i] := letters[ ctext[i] ]:
> od:
> evalm(ctext);
```

$$[F, K, X, J, E, T, J, M, Y, E, E, D, L, Q, K, N, A, A]$$

Thus, the resulting ciphertext is "FKXJETJMYEEDLQKNAA".

To decipher this message, we would begin by defining the ciphertext as the vector `ctext`, and defining `letters`, `ltable`, `A`, `blocksize`, and `numblocks` as before (defining `numblocks` as the number of ciphertext characters divided by `blocksize`). We would then need to convert the list of ciphertext letters into a list of ring elements. We can do this by entering the following commands.

```
> for i from 1 to vectdim(ctext) do
>       ctext[i] := ltable[ ctext[i] ]:
> od:
> evalm(ctext);
```

$$[5, 10, 23, 9, 4, 19, 9, 12, 24, 4, 4, 3, 11, 16, 10, 13, 0, 0]$$

Next, we would need to group these ciphertext ring elements into row vectors of length `blocksize`. We can do this by entering the following command, which leaves these row vectors in order as rows in the matrix `cmatrix`.

```
> cmatrix := matrix(numblocks, blocksize, ctext);
```

$$
cmatrix := \begin{bmatrix} 5 & 10 & 23 \\ 9 & 4 & 19 \\ 9 & 12 & 24 \\ 4 & 4 & 3 \\ 11 & 16 & 10 \\ 13 & 0 & 0 \end{bmatrix}
$$

We can then recover the plaintext ring elements by multiplying the preceding matrix by the inverse of the key matrix **A**. We can do this by entering the following command, which leaves the resulting product as the matrix `pmatrix`. (Note that to obtain the inverse of **A**, we must only raise **A** to the power -1.)

```
> pmatrix := map(m -> m mod 26, evalm(cmatrix &* A^(-1)));
```

$$
pmatrix := \begin{bmatrix} 17 & 4 & 13 \\ 3 & 4 & 25 \\ 21 & 14 & 20 \\ 18 & 0 & 19 \\ 13 & 14 & 14 \\ 13 & 0 & 0 \end{bmatrix}
$$

We can list the plaintext ring elements in order in the vector `ptext` by entering the following command.

```
> ptext := convert(pmatrix, vector);
```

$$
ptext := [\,17, 4, 13, 3, 4, 25, 21, 14, 20, 18, 0, 19, 13, 14, 14, 13, 0, 0\,]
$$

Finally, we can see the corresponding alphabet letters by entering the following commands.

```
> for i from 1 to vectdim(ptext) do
>     ptext[i] := letters[ ptext[i] ]:
> od:
> evalm(ptext);
```

$$
[\,R,\ E,\ N,\ D,\ E,\ Z,\ V,\ O,\ U,\ S,\ A,\ T,\ N,\ O,\ O,\ N,\ A,\ A\,]
$$

6.4 Generalizations of the Hill Cryptosystem

Just as we enhanced the security of our first encryption method in Section 6.1 by including a constant in the function $f : R \to R$, we can enhance the security of the Hill encryption method by including a vector of constants in the function $f : R^n \to R^n$. We summarize this as our next encryption method.

Generalized Hill Encryption Method: Let A be an $n \times n$ invertible matrix over R, and let B be a row vector of length n over R. Group the plaintext into row vectors P_i of length n, and define $f : R^n \to R^n$ by $f(P_i) = P_i A + B$ with each entry taken modulo $|R|$. The resulting rows listed together form the ciphertext.

The generalized Hill encryption method yields a system that has the matrix A and vector B as keys. The inclusion of B obviously yields a higher level of security than our original Hill encryption method. To break the generalized system by trial and error, an intruder would have to test a maximum of 26^{n^2+n} possible pairs (A, B) of keys, while to break the original Hill system an intruder would have to test a maximum of only 26^{n^2} possible keys. Even for very small values of n this increase is significant. For example, for $n = 5$, the generalized system has $26^5 = 11881376$ *times* as many possible pairs of keys as the number of possible keys for our original Hill system.

To decipher a message that has been encrypted using the generalized Hill encryption method with a key matrix A of size $n \times n$ and vector B of length n, we would group the ciphertext into row vectors C_i of length n, and compute $f^{-1}(C_i) = (C_i - B)A^{-1} \bmod |R|$ for all i. Hence, despite the somewhat significant increase in security yielded by the generalized Hill encryption method, there is not a significant increase in the computational work necessary for enciphering and deciphering messages.

We now discuss an extension of the generalized Hill encryption method that yields an even higher level of security.

Variable Matrix Encryption Method: Let A be an $n \times n$ invertible matrix over R, and let B_i be varying row vectors of length n over R. Group the plaintext into row vectors P_i of length n, and define $f_i : R^n \to R^n$ by $f_i(P_i) = P_i A + B_i$ with each entry taken modulo $|R|$. The resulting rows listed together form the ciphertext.

To decipher a message that has been encrypted using the variable matrix encryption method with a key matrix A of size $n \times n$ and vectors B_i of length n, we would group the ciphertext into row vectors C_i of length n, and compute $f_i^{-1}(C_i) = (C_i - B_i)A^{-1} \bmod |R|$ for all i.

One problem with the variable matrix encryption method is that in practice it may be difficult or cumbersome for the originator and intended recipient of the messages to keep a record of the vectors B_i. To avoid this problem, these vectors can be chosen so that they depend uniquely on the plaintext vectors P_i, the ciphertext vectors C_i, or the previous B_i. For example, three simple methods for choosing the B_i are

1. $B_i = P_{i-1}B$, where B is a fixed $n \times n$ matrix over R and P_0 is given,

2. $B_i = C_{i-1}B$, where B is a fixed $n \times n$ matrix over R and C_0 is given,

3. $B_i = (r_i, r_{i+1}, \ldots, r_{i+n-1})$, where $\{r_j\}$ is a recursive sequence over R with necessary initial r_j given.

Example 6.6 In this example we use the variable matrix encryption method to encipher the message "MEET AT SEVEN". We begin by converting the message into a list of elements in Z_{26}. The result of this conversion is shown at the beginning of Example 6.3. We will use the following key matrix A to encipher this message.

$$A = \begin{bmatrix} 1 & 2 & 1 \\ 3 & 1 & 0 \\ 0 & 2 & 1 \end{bmatrix}$$

And we will use the first of the three methods listed above for choosing the vectors B_i with the following matrix B

$$B = \begin{bmatrix} 1 & 0 & 0 \\ 1 & 1 & 1 \\ 0 & 0 & 1 \end{bmatrix}$$

and vector $P_0 = (1, 2, 3)$. To form the ciphertext, we group the plaintext into row vectors P_i of length 3, and compute $P_i A + B_i$ for all i with each entry taken modulo 26. Before constructing the first ciphertext vector, we construct the vector B_1 as follows.

$$B_1 = P_0 B = (1, 2, 3)B = (3, 2, 5)$$

We can then construct the first ciphertext vector as follows.

$$\begin{aligned} P_1 A + B_1 &= (12, 4, 4)A + (3, 2, 5) \\ &= (27, 38, 21) \\ &= (1, 12, 21) \end{aligned}$$

We construct the remaining vectors B_i as follows.

$$
\begin{aligned}
B_2 &= P_1 B &= (12, 4, 4)B &= (16, 4, 8) \\
B_3 &= P_2 B &= (19, 0, 19)B &= (19, 0, 19) \\
B_4 &= P_3 B &= (18, 4, 21)B &= (22, 4, 25)
\end{aligned}
$$

And we can then construct the remaining ciphertext vectors as follows.

$$
\begin{aligned}
P_2 A + B_2 &= (19, 0, 19)A + (16, 4, 8) &= \cdots &= (9, 2, 20) \\
P_3 A + B_3 &= (18, 4, 21)A + (19, 0, 19) &= \cdots &= (23, 4, 6) \\
P_4 A + B_4 &= (4, 13, 25)A + (22, 4, 25) &= \cdots &= (13, 23, 2)
\end{aligned}
$$

Hence, the entire encipherment is

	M	E	E	T	A	T	S	E	V	E	N	
$\alpha \Rightarrow$	12	4	4	19	0	19	18	4	21	4	13	25
$f \Rightarrow$	1	12	21	9	2	20	23	4	6	13	23	2
$\alpha^{-1} \Rightarrow$	B	M	V	J	C	U	X	E	G	N	X	C

and the ciphertext is "BMVJCUXEGNXC". ∎

6.5   The Two-Message Problem

Recall that the Hill cryptosystem can be used to obtain any desired level of security by simply choosing a key matrix that is sufficiently large. However, recall also that for the Hill system with an $n \times n$ key matrix, we discussed a technique in Section 6.2 by which an intruder may be able to break the system relatively easily if the intruder knows or is able to correctly guess n^2 characters from the plaintext. This technique illustrates the general fact that it is sometimes possible for an intruder to break a cryptosystem in an unusual way provided the intruder knows some additional information about the system (such as, for example, n^2 characters from the plaintext). In this section we discuss a technique by which an intruder can break a slight modification of the Hill system in an unusual way. We first discuss the modification of the system, which consists of using a key matrix that is *involutory*. A matrix K is said to be involutory if $K^2 = I$ (*i.e.*, if $K = K^{-1}$).

Modified Hill Encryption Method: Let K be an $n \times n$ involutory matrix over R. Group the plaintext into row vectors P_i of length n, and form ciphertext vectors C_i by $C_i = P_i K$ with each entry taken modulo $|R|$.

This modified Hill encryption method is in fact the method suggested by Hill when he first presented his cryptosystem in 1929. The reason for Hill's suggestion of using an involutory key matrix is obvious, for then the same matrix could be used to decipher a message that was used to encipher the message. Although this simplification of the Hill system is not as significant now with the recent developments in calculators and computers, the elimination of having to determine the inverse of the key matrix was very noteworthy in 1929. In fact, Hill even invented a machine designed to perform the calculations in his cryptosystem, and argued that by using an involutory key matrix one could both encipher and decipher a single message without "changing the settings".

Note that in order for the modified Hill encryption method to be a useful encryption technique, there should be a relatively large number of $n \times n$ involutory matrices for each $n > 1$. For any specific $n > 1$, if there are only a relatively few $n \times n$ involutory matrices, then an involutory key matrix of size $n \times n$ would certainly not yield a secure cryptosystem. We would assume in general that an intruder to such a system would know the size of the key matrix and the fact that the key matrix was involutory. Hence, if there are only a relatively few involutory matrices of size $n \times n$, an intruder could break the system very easily by simply testing each one. However, it is not the case that there are only a relatively few $n \times n$ involutory matrices for any $n > 1$. It has been well-known for many years that for any matrices A of size $r \times s$ and B of size $s \times r$ over a ring R, the block matrix

$$\begin{bmatrix} BA - I & B \\ 2A - ABA & I - AB \end{bmatrix}$$

is involutory over R (see Written Exercise 7). Thus, it is not unreasonable to suppose that an intruder should find the modified Hill system not significantly less difficult to break than the usual Hill system. But, as mentioned, it is sometimes possible for an intruder to break a cryptosystem in an unusual way provided the intruder knows some additional information about the system. As we discuss next, an intruder can break the modified Hill system in an unusual and relatively easy way provided the intruder intercepts two ciphertexts formed from the same plaintext using different involutory key matrices of the same size. In this scenario, the problem of breaking the system is called the *Two-Message* problem.

Suppose we intercept ciphertexts C, C' formed from the same plaintext P using the Hill cryptosystem with two different $n \times n$ involutory key matrices K, K'. That is, suppose a single plaintext P is grouped into row vectors

P_i of length n, and we intercept ciphertext vectors C_i and C_i' formed by

$$C_i = P_i K \qquad (6.4)$$

and

$$C_i' = P_i K' \qquad (6.5)$$

for all i where K, K' are distinct $n \times n$ involutory matrices. The Two-Message problem is to determine the plaintext vectors P_i for all i from knowledge of the ciphertext vectors C_i and C_i'. Note that since K and K' are involutory, they are their own corresponding decryption matrices. Thus, if we could determine K or K' we would be done. To do this, note that because K is involutory, (6.4) is equivalent to

$$P_i = C_i K \qquad (6.6)$$

for all i. By substituting this expression for P_i into (6.5), we obtain

$$C_i' = C_i K K' \qquad (6.7)$$

for all i. If there are n values of i, say $i_1, i_2, \ldots, i_n$, for which the $n \times n$ matrix $S = [C_{i_1}, C_{i_2}, \ldots, C_{i_n}]$ is invertible, then we can determine the matrix KK' as follows. Define the $n \times n$ matrix $T = [C_{i_1}', C_{i_2}', \ldots, C_{i_n}']$. Since (6.7) holds for every i, it follows that $T = SKK'$. Hence, we can determine KK' as $KK' = S^{-1}T$. Note that in order for the matrices S and T to exist, the message P must be at least n^2 characters in length. Note also that T will be invertible since T^{-1} can be expressed as $T^{-1} = K'KS^{-1}$.

Recall, as we stated above, if we can determine K or K', then the Two-Message problem will be solved. However, as we have just shown, with a very mild assumption we should not have any difficulty in finding the matrix KK'. After finding KK', we consider the matrix equation $(KK')X = X(KK')^{-1}$ or, equivalently,

$$(KK')X = X(K'K) \qquad (6.8)$$

for unknown matrix X. Note that both K and K' are involutory solutions to (6.8). Thus, if we find all of the involutory solutions to (6.8), the resulting collection of matrices will include both K and K'. To find the plaintext, we must then only decipher one of the ciphertexts with each of the involutory solutions to (6.8). Most likely only one of these decipherments, the correct plaintext, will make any sense. (To save time, we can decipher only a portion of one of the ciphertexts with each of the involutory solutions to (6.8). This will reveal which of the involutory solutions to (6.8) is the correct key matrix for that ciphertext. We can then use this key matrix to decipher the rest of the ciphertext.)

We now summarize the complete solution process for the Two-Message problem as follows.

1. Determine the matrix KK'.

2. Find all of the involutory solutions to (6.8).

3. Decipher one of the ciphertexts with each of the involutory solutions to (6.8). The correct key matrix, which is one of these involutory solutions, will yield the correct plaintext.

The procedure for completing the first step in this solution process was described previously, and can generally be done in a straightforward manner. The calculations for the third step can also generally be done in a straightforward manner, although due to the potentially large number of involutory solutions to (6.8), these calculations can be long and tedious. It is in the second step of this solution process that the essential difficulties of the Two-Message problem lie. This step can be considered in two parts. First, we can determine the general solution X to (6.8) by solving a system of n^2 linear equations for the unknown elements in X. After finding this general solution, we can find the involutory solutions to (6.8) by imposing the condition $X^2 = I$ on the general solution X. This requires solving a system of up to n^2 quadratic equations, hence providing many possible difficulties, especially for large n. However, the potential difficulties that can be incurred in solving a system of up to n^2 quadratic equations do not compare (time-wise) to those incurred in breaking the modified Hill system by using trial and error to determine the key matrix.

Written Exercises

1. Julius Caesar was known to encipher messages using our second encryption method from Section 6.1 with $a = 1$. This encryption method yields what is often called a *shift cipher*. Use a shift cipher to encipher the message "ATTACK AT DAWN". Also, describe and illustrate procedures the intended recipient could use to decipher the message and an intruder could use to break the system. Explain why it is natural to say that this encryption method yields a *shift* cipher.

2. Consider the following matrix A over Z_{26}.

$$A = \begin{bmatrix} 13 & 7 \\ 8 & 21 \end{bmatrix}$$

   (a) Use the Hill encryption method with the key matrix A defined above to encipher the message, "SEND TARGET STATUS".

   (b) Decipher "NDJLWLTBWFVXGSNV", a message that has been enciphered using the Hill encryption method with the key matrix A defined above.

3. Suppose you intercept "FLBIPURCRGAO", a message that has been enciphered using the Hill encryption method with a 2×2 key matrix A over Z_{26}, and you somehow know that the first four letters in the corresponding plaintext are "NCST". Decipher the message.

4. (a) Use the generalized Hill encryption method with the key matrix A from Written Exercise 2 and the vector $B = (1, 2)$ over Z_{26} to encipher the message, "ABORT MISSION".

   (b) Decipher "UXSJOEWNOJHE", a message that has been enciphered using the generalized Hill encryption method with the key matrix A from Written Exercise 2 and the vector $B = (1, 2)$ over Z_{26}.

5. Consider the following matrices A and B and vector C_0 over Z_{26}.

$$A = \begin{bmatrix} 2 & 5 \\ 1 & 4 \end{bmatrix} \qquad B = \begin{bmatrix} 1 & 0 \\ 1 & 1 \end{bmatrix} \qquad C_0 = (1, 2)$$

   (a) Use the variable matrix encryption method with the key matrix A defined above to encipher the message, "NEED BACKUP". For choosing the vectors B_i, use the second of the three methods listed immediately before Example 6.6 with the matrix B and vector C_0 defined above.

(b) Decipher "RTOWKRPTLS", a message that has been enciphered using the variable matrix encryption method with the key matrix A defined above. The vectors B_i were chosen using the second of the three methods listed immediately before Example 6.6 with the matrix B and vector C_0 defined above.

6. Consider the recursive sequence $\{r_j\}$ over Z_{26} given by

$$r_{j+2} = (r_{j+1} + r_j) \bmod 26$$

with $r_1 = 3$ and $r_2 = 5$.

(a) Use the variable matrix encryption method with the key matrix A from Written Exercise 5 to encipher the message, "GO PACK". For choosing the vectors B_i, use the third of the three methods listed immediately before Example 6.6 with the recursive sequence $\{r_j\}$ defined above.

(b) Decipher "JAYGKI", a message that has been enciphered using the variable matrix encryption method with the key matrix A from Written Exercise 5. The vectors B_i were chosen using the third of the three methods listed immediately before Example 6.6 with the recursive sequence $\{r_j\}$ defined above.

7. Consider matrices A of size $r \times s$ and B of size $s \times r$ over a ring R. Find the size of the matrix

$$\begin{bmatrix} BA - I & B \\ 2A - ABA & I - AB \end{bmatrix}$$

and show that this matrix is involutory over R.

8. Use the result from Written Exercise 7 to construct an involutory matrix of size 3×3 over Z_{26}. Then use your result as the key matrix K in the modified Hill encryption method to encipher a plaintext of your choice with at least six characters. Also, show how to decipher the resulting ciphertext.

9. Let A be a matrix of size $n \times n$ over Z_k. Recall that

$$A \,(\mathrm{adj}\, A) = (\mathrm{adj}\, A)\, A = (\det A)\, I$$

where $\mathrm{adj}\, A$ represents the adjoint of A and I is the $n \times n$ identity. Show that A is invertible over Z_k if and only if the determinant of A has a multiplicative inverse in Z_k.

10. Show that $a \in Z_k$ has a multiplicative inverse in Z_k if and only if $(a, k) = 1$.

Maple Exercises

1. Consider the following matrix A over Z_{26}.

$$A = \begin{bmatrix} 9 & 10 & 0 & 20 & 7 \\ 4 & 3 & 14 & 23 & 16 \\ 7 & 2 & 5 & 7 & 5 \\ 21 & 1 & 25 & 3 & 1 \\ 1 & 5 & 4 & 3 & 0 \end{bmatrix}$$

(a) Use the Hill encryption method with the key matrix A defined above to encipher the message, "ABORT MISSION PROCEED WITH SECONDARY ORDERS".

(b) Decipher "ZQBTDDBIGZZDCQFRXFBXPVJERZRSBA", a message that has been enciphered using the Hill encryption method with the key matrix A defined above.

2. Consider the following matrix B and vector P_0 over Z_{26}.

$$B = \begin{bmatrix} 13 & 22 & 4 & 4 & 3 \\ 2 & 0 & 4 & 6 & 8 \\ 1 & 25 & 17 & 23 & 9 \\ 3 & 2 & 6 & 3 & 12 \\ 7 & 4 & 5 & 3 & 12 \end{bmatrix} \qquad P_0 = (1,1,1,1,1)$$

(a) Use the variable matrix encryption method with the key matrix A from Maple Exercise 1 to encipher the message, "ATTACK FLANK AT SUNRISE". For choosing the vectors B_i, use the first of the three methods listed immediately before Example 6.6 with the matrix B and vector P_0 defined above.

(b) Decipher "ZRLGVCKZHWLMOSHXOGBU", a message that has been enciphered using the variable matrix encryption method with the key matrix A from Maple Exercise 1. The vectors B_i were chosen using the first of the three methods listed immediately before Example 6.6 with the matrix B and vector P_0 defined above.

3. Use the result from Written Exercise 7 to construct an involutory matrix of size 5×5 over Z_{26}. (You may find the Maple **blockmatrix** command useful in constructing this matrix.) Then use your result as the key matrix K in the modified Hill encryption method to encipher a plaintext of your choice with at least 15 characters. Also, show how to decipher the resulting ciphertext.

Chapter 7

The RSA Cryptosystem

In this chapter, we discuss one of the most well-known and popular cryptosystems ever developed – the *RSA* cryptosystem. One reason why it is so well-known and popular is that it is a classic public-key system. Recall in general that everything in a cryptosystem is assumed to be public knowledge except the parameters in the enciphering function. A public-key system is one in which even these parameters can be public knowledge without compromising the security of the system. That is, using the notation introduced in Section 6.1, a public-key system is one in which the function f can be public knowledge without revealing f^{-1}. The RSA cryptosystem is named for R. Rivest, A. Shamir, and L. Adleman, who first published it in 1978.

Before formally presenting the RSA system, we consider a very simple example of the mathematics that govern it. Choose primes $p = 5$ and $q = 11$, and let $n = pq = 55$ and $m = (p-1)(q-1) = 40$. Next, choose $a = 27$, chosen so that $(a, m) = 1$, and let $b = 3$, chosen so that $ab = 1 \bmod m$. Then for $x = 2$, note that

$$(x^a)^b = (2^{27})^3 = 2417851639229258349412352 = 2 \bmod 55 = x \bmod n.$$

An important thing to note about this computation is that

$$x^{ab} = x \bmod n. \tag{7.1}$$

In fact, this equation will be true for any $x \in Z$ because a and b were chosen so that

$$ab = 1 \bmod m. \tag{7.2}$$

Thus, if we encipher a message by raising the plaintext to the power a, we can decipher the message by raising the ciphertext to the power b and reducing modulo n. It is certainly not obvious that (7.1) will hold for any $x \in Z$ provided (7.2) is true. Establishing this result will be one of our primary goals in Section 7.1.

7.1   Mathematical Prerequisites

Before establishing the fact that (7.1) will hold for any $x \in Z$ provided (7.2) is true, we discuss some additional preliminaries. We first discuss how to find values of a and b that satisfy (7.2). Of course, it is not difficult to choose a, as it must only be relatively prime to m. Once a is chosen, we can then find b as the multiplicative inverse of a modulo m by constructing the Euclidean algorithm table (see Section 1.6) for a and m. We illustrate this in the following example.

Example 7.1 Consider $a = 27$ and $m = 40$. To find a value for b that satisfies $ab = 1 \bmod m$, we first apply the Euclidean algorithm to a and m as follows.

$$
\begin{aligned}
40 &= 27 \cdot 1 + 13 \\
27 &= 13 \cdot 2 + 1
\end{aligned}
$$

Note that, as required, $(a, m) = 1$. Constructing the Euclidean algorithm table for a and m, we obtain the following.

Row	Q	R	U	V
-1	$-$	40	1	0
0	$-$	27	0	1
1	1	13	1	-1
2	2	1	-2	3

Hence, $40(-2) + 27(3) = 1$. This immediately gives the result, for it states that $27(3) = 1 \bmod 40$, and thus $b = 3$ satisfies $ab = 1 \bmod m$. ∎

Next, we show the general relationship between the values of n and m in the example in the introduction to this chapter. To do this we must first prove some general results about the ring of integers.

Let n be an integer with $n > 1$. Then the modular ring Z_n inherits many properties from Z since it is a quotient ring of Z. Consider the set U_n of units in Z_n. That is, consider the set $U_n = \{x \in Z_n \mid x \text{ has a multiplicative inverse in } Z_n\}$. Note that U_n can also be expressed as $U_n = \{x \in Z_n \mid (x, n) = 1\}$, and that U_n forms a multiplicative group (see Written Exercise 11). The order of U_n is denoted in general by $\varphi(n)$. The function φ is called the *Euler-phi* function.

Theorem 7.1 *If p is a prime, then $U_p = Z_p^*$ and $\varphi(p) = p - 1$.*

Proof. Exercise. ∎

Theorem 7.2 *Suppose a and b are integers with $(a, b) = 1$. Then $\varphi(ab) = \varphi(a)\varphi(b)$.*

Proof. Consider the sets $A = \{s \mid 0 < s \leq a$ and $(s, a) = 1\}$ and $B = \{r \mid 0 < r \leq b$ and $(r, b) = 1\}$. Then $|A| = \varphi(a)$ and $|B| = \varphi(b)$. Now, let $T = \{ar + bs \mid r \in B$ and $s \in A\}$. We claim that $(ar + bs, ab) = 1$. Suppose there exists a prime p that divides both $ar + bs$ and ab. Then $p|a$ or $p|b$. If $p|a$, then $p|ar + bs$ implies $p|bs$. But since $(a, b) = 1$, then p does not divide b. Hence, $p|s$. But $(s, a) = 1$, which is a contradiction. A similar argument holds if $p|b$. Thus, $(ar + bs, ab) = 1$. Suppose now that $ar + bs = ar' + bs'$ mod ab with $r, r' \in B$ and $s, s' \in A$. Then $a(r - r') = b(s' - s)$ mod ab, and $b|a(r - r')$. Hence, $b|(r - r')$ since $(a, b) = 1$. But $0 < r$ and $r' \leq b$. Therefore, $r = r'$. Similarly, $s = s'$. Let $U = \{w \mid w$ is the remainder when $ar + bs$ is divided by $ab\}$. Each element in U is relatively prime to ab, and $|U| = |A \times B|$. Hence, $\varphi(ab) \geq |U| = |A \times B| = \varphi(a)\varphi(b)$. To show the desired equality, it is now sufficient to show that if $c \in Z$ with $(c, ab) = 1$ and $0 < c \leq ab$, then $c \in U$. Since $(a, b) = 1$ and $ax + by = 1$ for some $x, y \in Z$, then $axc + byc = c$. Let $z = xc$ and $t = yc$. Then $az + bt = c$. Since $(a, c) = 1$, it then follows that $(a, t) = 1$. Thus, $t = s$ mod a for some $s \in A$, and $bt = bs$ mod ab. In a similar manner it can be seen that $az = ar$ mod ab for some $r \in B$. Hence, $ar + bs = c$ mod ab, and $c \in U$. ∎

The reason Theorems 7.1 and 7.2 are of interest to us is because of the following corollary, which states the general relationship between the values of n and m in the example in the introduction to this chapter.

Corollary 7.3 *Suppose $n = pq$ for distinct primes p and q. Then $\varphi(n) = (p - 1)(q - 1)$.*

Proof. Exercise. ∎

We now show that (7.1) will hold for any $x \in Z$ provided (7.2) is true. The main result we will use to show this is the following theorem, commonly called *Fermat's Little Theorem*.

Theorem 7.4 *Let p be a prime, and suppose $x \in Z$ satisfies $(x, p) = 1$.
Then $x^{p-1} = 1 \bmod p$.*

Proof. We claim first that the elements in the set

$$S_1 = \{x \bmod p, 2x \bmod p, \ldots, (p-1)x \bmod p\}$$

are a rearrangement of the elements in the set

$$S_2 = \{1, 2, \ldots, p-1\}.$$

To see this, note that if $jx \bmod p = kx \bmod p$ for some positive integers j
and k less than p, then $p|(j-k)x$. But since p does not divide x, this implies
that $p|j - k$. Thus, because j and k are less than p, it follows that $j = k$.
Now, since $S_1 = S_2$, the product of the elements in S_1 will be equal to the
product of the elements in S_2. That is, $x^{p-1}(p-1)! \bmod p = (p-1)! \bmod p$.
Hence, $p|(p-1)!(x^{p-1} - 1)$. Finally, since p does not divide $(p-1)!$, then
$p|x^{p-1} - 1$ or, equivalently, $x^{p-1} = 1 \bmod p$.   ∎

In the following theorem we establish the fact that (7.1) will hold
provided (7.2) is true.

Theorem 7.5 *Suppose p and q are distinct primes and define $n = pq$ and
$m = \varphi(n) = (p-1)(q-1)$. If a and b are integers that satisfy $ab = 1 \bmod m$,
then $x^{ab} = x \bmod n$ for all $x \in Z$.*

Proof. Since $ab = 1 \bmod m$, then $ab = 1 + km$ for some $k \in Z$. Hence, for
any $x \in Z$, it follows that

$$x^{ab} = x^{1+km} = x(x^{km}) = x(x^{p-1})^{k(q-1)}.$$

If $(x, p) = 1$, then by Theorem 7.4 we know that $x^{p-1} = 1 \bmod p$. Thus,
$x^{ab} = x(1)^{k(q-1)} \bmod p = x \bmod p$. Also, if $(x, p) \neq 1$, then $x = 0 \bmod p$,
and certainly $x^{ab} = x \bmod p$. Similarly, $x^{ab} = x \bmod q$. Hence, $p|(x^{ab} - x)$
and $q|(x^{ab} - x)$, and thus $pq|(x^{ab} - x)$. That is, $n|(x^{ab} - x)$ or, equivalently,
$x^{ab} = x \bmod n$.   ∎

7.2   RSA Encryption and Decryption

To encipher a message using the RSA cryptosystem, we first convert the
message into a list of nonnegative integers by applying a mapping like the

correspondence α from Chapter 6. We then choose distinct primes p and q and define $n = pq$ and $m = \varphi(n) = (p-1)(q-1)$. Next, we choose $a \in Z_m^*$ such that $(a, m) = 1$, and find $b \in Z_m^*$ that satisfies $ab = 1 \bmod m$. (Recall that b can be found by constructing the Euclidean algorithm table for a and m.) To encipher the message, we form ciphertext integers by raising the plaintext integers to the power a and reducing modulo n. According to Theorem 7.5, we can then recover the plaintext integers by raising the ciphertext integers to the power b and reducing modulo n.

Example 7.2 In this example, we use the RSA cryptosystem to encipher and decipher the message, "NCSU". We first apply the correspondence α from Chapter 6 to convert this message into the list of integers 13 2 18 20. Next, we choose primes $p = 5$ and $q = 11$, and define $n = pq = 55$ and $m = (p-1)(q-1) = 40$. We then choose encryption exponent $a = 27$. To encipher the message, we perform the following calculations.

$$
\begin{aligned}
13^{27} &= 7 \bmod 55 \\
2^{27} &= 18 \bmod 55 \\
18^{27} &= 17 \bmod 55 \\
20^{27} &= 15 \bmod 55
\end{aligned}
$$

Hence, the ciphertext is the list of integers 7 18 17 15. (Although we could use α^{-1} to convert this particular list of integers back into a list of letters, conversion of an RSA ciphertext back into a list of alphabet characters is not usually possible. To see this, note that because the results of these encryption calculations were reduced modulo $n = 55$, these results could have been as large as $n - 1 = 54$.) By Example 7.1, the decryption exponent that corresponds to the encryption exponent $a = 27$ in this example is $b = 3$. Hence, to decipher the message, we must only perform the following calculations.

$$
\begin{aligned}
7^3 &= 13 \bmod 55 \\
18^3 &= 2 \bmod 55 \\
17^3 &= 18 \bmod 55 \\
15^3 &= 20 \bmod 55
\end{aligned}
$$

Note that the results are the original plaintext integers.　■

We still have several topics to address regarding the RSA cryptosystem. Note first that no matter how large we choose the encryption exponent and modulus for the RSA system, the system as illustrated in Example 7.2 will certainly not be secure because it will just yield a substitution cipher. However, we can use the RSA encryption procedure as presented to obtain a non-substitution cipher by simply grouping consecutive integers in the plaintext before enciphering. Because our exponentiation operations are done modulo n, we will still be able to convert between plaintext and ciphertext uniquely, provided the plaintext integers are grouped into blocks that are smaller than n. We illustrate this in the following example.

Example 7.3 In this example, we again encipher and decipher the message, "NCSU". We begin by choosing primes $p = 79$ and $q = 151$, and defining $n = pq = 11929$ and $m = (p - 1)(q - 1) = 11700$. Next, we choose $a = 473$, chosen so that $(a, m) = 1$. We can then use the Euclidean algorithm to find that $b = 8237$ satisfies $ab = 1 \bmod m$. Recall that our plaintext converts to the list of integers 13 2 18 20. Since we have chosen a 5-digit value for n, we can group the first two and last two plaintext integers into blocks that will be smaller than n. That is, we can express the plaintext as 1302 1820 (note that we use 02 for 2), and use the RSA encryption procedure as presented. To encipher the message, we perform the following calculations.

$$1302^{473} = 7490 \bmod 11929$$
$$1820^{473} = 9723 \bmod 11929$$

Hence, the ciphertext is 7490 9723. To decipher the message, we perform the following calculations.

$$7490^{8237} = 1302 \bmod 11929$$
$$9723^{8237} = 1820 \bmod 11929$$

We can then split the resulting 4-digit integers into the original 2-digit plaintext integers. ∎

Another topic we must address regarding the RSA cryptosystem is how the system actually progresses between two people wishing to exchange a secret message. We stated in the introduction to this chapter that the RSA cryptosystem is a public-key system. This forces the system to progress in a particular way.

Recall that in general we assume almost everything in a cryptosystem is public knowledge, including the form of the enciphering function. This means that we would assume an intruder who intercepts an RSA ciphertext would know that each ciphertext integer was formed as $x^a \bmod n$ for some plaintext integer x and positive integers a and n. The fact that the RSA cryptosystem is a public-key system means that we would assume the intruder also knows the actual values of a and n used in the encryption. For example, we would assume an intruder who intercepts the ciphertext in Example 7.3 would know that each ciphertext integer was formed as $x^{473} \bmod 11929$ for some plaintext integer x. Although this obviously affects the security of the system, we make this assumption because in practice the RSA system is used with a and n being public knowledge. The benefit of this is that two people wishing to use RSA to send a secret message across an insecure line of communication do not have to figure out a way to secretly exchange an encryption exponent and modulus.

The comments made in the previous paragraph imply that the RSA system in Example 7.3 is not mathematically secure. This is because an intruder could mathematically break the system as follows. After finding the values of p and q in $n = pq = 11929$, an intruder could form $m = (p-1)(q-1)$, use the Euclidean algorithm to find that $b = 8237$ satisfies $ab = 1 \bmod m$, and decipher the message by raising the ciphertext integers to the power b and reducing modulo n. Hence, none of the operations necessary to break this system would take an intruder more than a few minutes. And even with significantly larger numbers, the Euclidean algorithm and modular exponentiation can easily be efficiently programmed on a computer. However, the first step in this process requires an intruder to find the two prime factors of n. It is the apparent difficulty of this problem, provided p and q are very large, that gives the RSA system an extremely high level of security. For example, if p and q are both around 100 digits long, the fastest known factoring algorithms would generally take millions of years to factor $n = pq$, even when programmed on a computer that can perform millions of operations per second. (We make some comments on choosing very large prime numbers in Sections 7.3 and 7.5, and some comments on factoring numbers with very large prime factors in Sections 7.3 and 7.6.) Hence, even if the encryption exponent a is public knowledge, an intruder should not be able to determine the decryption exponent b. This is precisely why the RSA cryptosystem is called a *public-key* system – the parameters in the enciphering function $f(x) = x^a \bmod n$ can be public knowledge without revealing the parameter b in the deciphering function $f^{-1}(x) = x^b \bmod n$.

We now mention how the RSA system actually progresses between two people wishing to exchange a secret message across an insecure line of communication. Because only the intended recipient of the message must be able to decipher the message, the intended recipient of the message initiates the process by choosing primes p and q, and defining $n = pq$ and $m = (p-1)(q-1)$. The intended recipient then chooses an encryption exponent $a \in Z_m^*$ such that $(a, m) = 1$ and, using the Euclidean algorithm if necessary, finds $b \in Z_m^*$ that satisfies $ab = 1 \bmod m$. The intended recipient then sends the values of a and n to the originator of the message across the insecure line of communication, forcing the assumption that a and n are public knowledge. The originator of the message enciphers the message by applying the function $f(x) = x^a \bmod n$ to the plaintext integers, and then sends the resulting ciphertext integers to the intended recipient across the insecure line of communication. Since only the intended recipient knows b, only the intended recipient can decipher the message by applying the function $f^{-1}(x) = x^b \bmod n$ to the ciphertext integers.

Example 7.4 Suppose we wish to use the RSA cryptosystem to send the message, "NCSU" to a colleague across an insecure line of communication. Our colleague begins the process by choosing primes p and q, and defining $n = pq = 363794227$ and $m = (p-1)(q-1)$. Next, our colleague chooses $a = 13783$, chosen so that $(a, m) = 1$, and uses the Euclidean algorithm to find $b \in Z_m^*$ that satisfies $ab = 1 \bmod m$. Our colleague then sends the values of a and n to us across the insecure line of communication. Recall that our plaintext converts to the list of integers   13   02   18   20. Since our colleague has chosen a 9-digit value for n, we can group all four of these 2-digit plaintext integers into a single block that will be smaller than n. That is, we can express the plaintext as   13021820, and encipher our message by applying the function $f(x) = x^a \bmod n$ to this plaintext integer. To encipher the message we perform the following calculation.

$$13021820^{13783} \quad = \quad 91518013 \bmod 363794227$$

We would then transmit the ciphertext integer 91518013 to our colleague across the insecure line of communication. In order for an intruder who intercepts this ciphertext and the previously transmitted values of a and n to decipher the message, the intruder would need to find the decryption exponent b. But to find b, an intruder would first need to find m. And to find m, an intruder would need to find the prime factors of n, a problem that, as we have stated, is essentially impossible provided our colleague has chosen sufficiently large values for p and q. This would not pose a problem for our colleague, however, because our colleague began the process by choosing p and q. Hence, our colleague would know that the prime factors of $n = pq = 363794227$ are $p = 14753$ and $q = 24659$, and would have no difficulty in forming $m = (p-1)(q-1) = 363754816$ and using the Euclidean algorithm to find that $b = 20981287$ satisfies $ab = 1 \bmod m$. To decipher the message, our colleague would then only need to perform the following calculation.

$$91518013^{20981287} \quad = \quad 13021820 \bmod 363794227$$

(We make some comments on efficiently raising large numbers to large powers in Sections 7.3 and 7.4.)   ∎

7.3 The RSA Cryptosystem with Maple

In this section we show how Maple can be used to encipher and decipher messages using the RSA cryptosystem.

We begin by mentioning several Maple commands that are useful for finding large primes. The first command we will mention is the **nextprime** command, which returns the smallest prime larger than an integer input. For example, the following command returns the smallest prime larger than 400043344212007458000.

> nextprime(400043344212007458000);

$$400043344212007458013$$

A similar command is the **prevprime** command, which returns the largest prime smaller than an integer input. For example, the following command returns the largest prime smaller than 400043344212007458000.

> prevprime(400043344212007458000);

$$400043344212007457977$$

A final primality command we will mention is the **isprime** command, which returns *true* if an integer input is prime and *false* if not. For example, the following commands imply that 400043344212007457977 is prime while 400043344212007458000 is not.

> isprime(400043344212007457977);

$$true$$

> isprime(400043344212007458000);

$$false$$

We should mention that the **nextprime, prevprime,** and **isprime** commands are probabilistic routines that employ primality tests (see Section 7.5). This means that the output returned by Maple is in general guaranteed to be correct with extremely high probability, but not absolutely.

We now show how Maple can be used to perform the RSA encipherment and decipherment procedures. We begin by finding large primes p and q.

> p := nextprime(400043344212007458000);

$$p := 400043344212007458013$$

> q := nextprime(500030066366269001200);

$$q := 500030066366269001203$$

Next, we define $n = pq$ and $m = (p-1)(q-1)$.

```
> n := p*q;
```

$$n := 2000336999557142833451725215840084689896639$$

```
> m := (p-1)*(q-1);
```

$$m := 2000336999557142833442724481734301925 30424$$

And we will use the following encryption exponent a.

```
> a := 100987689009876790009100 03;
```

$$a := 100987689009876790009100 03$$

To verify that this value of a is a valid RSA encryption exponent, we enter the following Maple **igcd** command, which returns the greatest common divisor of the integers a and m. Note that, as required, $(a, m) = 1$.

```
> igcd(a, m);
```

$$1$$

We now use the RSA encipherment procedure to encipher the message, "RETURN TO HEADQUARTERS". (Because the letter "I" represents $\sqrt{-1}$ in Maple, the user-written procedures that follow had to be designed for messages expressed with lower-case letters. Also, contrary to the way we defined messages in Section 6.3, note that the following message is defined as a string of letters without spaces rather than as a vector containing the letters.)

```
> message := 'returntoheadquarters';
```

$$message := returntoheadquarters$$

Next, we convert this message into a list of 2-digit integers and combine these integers into a single block. To do this, we have provided the user-written procedure **to_number**, for which code is given in Appendix C.2. If this procedure is saved as the text file *to_number* in the directory from which we are running Maple, then we can include the **to_number** procedure in this Maple session by entering the following command.

```
> read to_number;
```

We can then convert **message** into its numerical equivalent as a single block by entering the following command.

```
> plaintext := to_number(message);
```

$$plaintext := 1704192017131914070400031620001719041718$$

Because this plaintext integer is smaller than n, we can encipher this message as a single block. That is, we can encipher this message by raising **plaintext** to the power a and reducing modulo n. To do this, we enter the following command. (Because this modular exponentiation involves such a large exponent, we use the Maple **&^** command instead of just **^** for the exponentiation. By using **&^**, we cause Maple to do the exponentiation in a very efficient way, like the technique discussed in Section 7.4.)

```
> ciphertext := plaintext &^ a mod n;
```

$ciphertext := 39705667751051336812284136334817473485289$

To decipher this message, we must find a decryption exponent b that satisfies $ab = 1 \bmod m$. We can do this by entering the following Maple **igcdex** command. Like the preceding **igcd** command, the following **igcdex** command returns the greatest common divisor of the integers a and m. However, the following **igcdex** command also takes two additional user-defined variable inputs, which it leaves as integers b and y that satisfy $ab + my = (a, m)$. Since $(a, m) = 1$, these will be values of b and y that satisfy $ab + my = 1$ or, equivalently, $ab = 1 \bmod m$. Thus, we can find a decryption exponent b by entering the following command.

```
> igcdex(a, m, 'b', 'y');
```

$$1$$

To see the decryption exponent b defined by the previous command, and to express this value as a positive number less than m, we enter the following command.

```
> b := b mod m;
```

$b := 54299300950841826990071853678997985400035$

Next, by entering the following command we verify that this value of b satisfies $ab = 1 \bmod m$.

```
> a*b mod m;
```

$$1$$

To recover the plaintext integer, we must only raise **ciphertext** to the power b and reduce modulo n.

```
> plaintext := ciphertext &^ b mod n;
```

$plaintext := 1704192017131914070400031620001719041718$

To see the original plaintext letters, we must split this single block into a list of 2-digit integers and convert these integers back into letters. To do this, we have provided the user-written procedure **to_letter**, for which code is given in Appendix C.2. If this procedure is saved as the text file *to_letter* in the directory from which we are running Maple, then we can include the **to_letter** procedure in this Maple session by entering the following command.

> `read to_letter;`

We can then convert **plaintext** back into a list of letters by entering the following command.

> `to_letter(plaintext);`

$$returntoheadquarters$$

A final command we will mention is the **ifactor** command, which returns the prime factorization of an integer input. For example, the following command very quickly returns the prime factorization of the 43-digit integer 1118516508138307725195354324934560155358253.

> `ifactor(1118516508138307725195354324934560155358253);`

$$(17)^5 (389) (45001200019828331)^2$$

Recall that the security of the RSA cryptosystem is based on the apparent difficulty of factoring the value of n. Hence, in order for the RSA system used in this section to be secure, it should be very difficult for an intruder to factor the 42-digit value of n used in this section. Although this value of n is one digit shorter than the integer used in the preceding command, because the prime factors of n are both very large, it will take **ifactor** much more time to return these prime factors. For example, the reader may wish to enter the preceding and following commands to see the difference. (Make sure you know how to interrupt the following command before entering it.)

> `ifactor(200033699955714283345172521584008468989639);`

And recall, as mentioned in Section 7.2, if p and q are both around 100 digits long, then the fastest known factoring algorithms, including the one employed by the **ifactor** command, would in general take millions of years to factor $n = pq$, even when programmed on a computer that can perform millions of operations per second.

7.4 A Note on Modular Exponentiation

Securely enciphering and deciphering messages using the RSA cryptosystem generally requires modular exponentiation with extremely large bases and exponents. For example, to decipher the message in Section 7.3, we had to raise the number 397056677510513368122841363348174734853289 to the power 542993009508418269900718536789979854000035 and reduce the result modulo 2000336999557142833451725215840084689899639. Even using the world's fastest computer, performing this computation by actually multiplying 397056677510513368122841363348174734853485289 by itself 542993009508418269900718536789979854000034 times would take essentially an infinite amount of time. In this section we show a technique that can be used to perform this modular exponentiation in a very efficient way.

For convenience, we will illustrate this technique for efficient modular exponentiation in the calculation

$$91518013^{20981287} = 13021820 \bmod 363794227 \qquad (7.3)$$

that deciphered the message in Example 7.4. This modular exponentiation can be done in a much more efficient way than multiplying 91518013 by itself 20981286 times. To do this computation more efficiently, we begin by computing the values of $(91518013)^{2^i} \bmod 363794227$ for $i = 1, \ldots, 24$. That is, for $P = 91518013$ and $M = 363794227$, we compute $P^2, P^4, P^8, P^{16}, \ldots, P^{2^{24}}$, and reduce each modulo M. Note that each $P^{2^i} \bmod M$ can be found by squaring $P^{2^{i-1}} \bmod M$. Thus, finding these values requires 24 total multiplications. The modular exponentiation in (7.3) can then be completed by computing

$P^{20981287} \bmod M$

$$
\begin{aligned}
&= P^{16777216+4194304+8192+1024+512+32+4+2+1} \bmod M \\
&= P^{2^{24}+2^{22}+2^{13}+2^{10}+2^9+2^5+2^2+2^1+2^0} \bmod M \\
&= P^{2^{24}} \cdot P^{2^{22}} \cdot P^{2^{13}} \cdot P^{2^{10}} \cdot P^{2^9} \cdot P^{2^5} \cdot P^{2^2} \cdot P^{2^1} \cdot P^{2^0} \bmod M
\end{aligned}
$$

which requires only 8 additional multiplications. Hence, this technique can be used to perform the modular exponentiation in (7.3) with only 32 multiplications. This is, of course, much fewer than the 20981286 multiplications necessary to multiply P by itself 20981286 times.

It is not difficult to see that this technique for efficiently computing $P^a \bmod M$ requires at most $2 \log_2 a$ multiplications (see Written Exercise 6). Hence, to perform even the massive modular exponentiation mentioned at the beginning of this section, this technique would require at most only

$$2 \log_2 542993009508418269900718536789979854000035 \approx 270$$

multiplications.

7.5   A Note on Primality Testing

Recall that to construct a secure RSA cryptosystem, the primes p and q chosen for the system must be very large. For example, we mentioned in Section 7.2 that if p and q are both around 100 digits long, it would in general take an intruder millions of years to break the system, even using a computer that can perform millions of operations per second. However, constructing an RSA system with such large primes is not particularly easy, for it is not particularly easy to find such large primes. In fact, motivated in part by the development of public-key cryptosystems like the RSA system, much research has been done recently in the area of primality testing.

Contrary to what the general name *primality* test suggests, a primality test is a criterion that can be used to determine if a specific number is *not* prime. The conclusions that can be drawn from applying a primality test to a number n are that either n "fails" the test and is definitely not prime, or that n "passes" the test and is probably prime (with probability depending on the "power" of the test).

The most direct and conclusive way to determine if a large odd integer n is prime is to try to find nontrivial factors of n by trial and error. We can do this systematically by checking to see if $m|n$ as m runs through the odd integers starting with $m = 3$ and stopping when m reaches $\sqrt{n}$. While this would reveal with certainty whether n was prime or composite, it would require many more divisions than could reasonably be done if n was of any significant size. In the remainder of this section we briefly discuss a very well-known and simple primality test based on Fermat's Little Theorem (Theorem 7.4).

If n is prime, then as a consequence of Fermat's Little Theorem,

$$a^{n-1} = 1 \bmod n \qquad (7.4)$$

for all $a \in Z_n^*$. Hence, it follows that if $a^{n-1} \neq 1 \bmod n$ for any $a \in Z_n^*$, then n is not prime. Thus, we can test the primality of an integer n by checking to see if (7.4) holds for certain values of a in Z_n^*. While this test is very easy to perform, there are some values of a for which (7.4) holds even when $(a, n) = 1$ and n is composite. In such cases, n is called a *pseudoprime* to the base a. For example, $2^{340} = 1 \bmod 341$ even though 341 is not prime. Thus, 341 is a pseudoprime to the base 2. However, since $3^{340} = 56 \bmod 341$, then 341 is not a pseudoprime to the base 3.

Pseudoprimes are scarce relative to the primes. For example, there are only 245 pseudoprimes to the base 2 less than one million, while there are 78498 primes less than one million. Also, most pseudoprimes to the base 2 are not pseudoprimes to many other bases. However, there do exist composite integers n that are pseudoprime to every base $a < n$ with $(a, n) = 1$. Such numbers are called *Carmichael* numbers. There are 2163 Carmichael numbers less than 2.5×10^{10}. The smallest Carmichael number is 561.

There are many primality tests that are more definitive in their conclusions than the test described above. For example, a further primality test based on Fermat's Little Theorem that is also very easy to perform fails only for an extremely small number of composites called *strong* pseudoprimes. In fact, there is only one strong pseudoprime to the bases 2, 3, 5, and 7 less than 2.5×10^{10}. There is no strong pseudoprime analogue to Carmichael numbers.

7.6 A Note on Integer Factorization

Recall that the security of the RSA cryptosystem is based on the apparent difficulty of factoring a number that is the product of two very large distinct primes. As in the area of primality testing, the development of public-key cryptosystems like the RSA system has motivated much research in the area of integer factorization. In this section we briefly discuss a very simple technique for integer factorization called *Fermat factorization*. Despite the fact that this factorization technique is quite old, Fermat factorization is still a very useful technique for factoring numbers that are the product of two very large distinct primes that are relatively close together.

Let $n = pq$ be the product of two very large distinct primes, and suppose we wish to determine the values of p and q from the knowledge of n. The most direct way to find p and q would be by trial and error. However, this would certainly not be feasible if both p and q were of any significant size. But if p and q were relatively close together, then even if they were very large we could determine them as follows. Let $x = (p+q)/2$ and $y = (p-q)/2$. Then $n = pq = x^2 - y^2 = (x+y)(x-y)$. And since n has prime factors p and q, then p and q must be equal to $x+y$ and $x-y$. Hence, to determine p and q, we must only find the values of x and y. To find x and y, we begin by assuming that x is the smallest integer larger than $\sqrt{n}$. Since $n = x^2 - y^2$, if we have assumed the correct value of x, then $x^2 - n$ will be the perfect square y^2. If $x^2 - n$ is not a perfect square, then we have assumed an incorrect value for x, and we increase x by one and repeat. We continue to repeat this process, each time increasing x by one, until $x^2 - n$ is a perfect square. Note that if p and q are relatively close together, then the number of times this process must be repeated should be relatively small.

Example 7.5 Suppose that we wish to find the two prime factors of $n = pq = 64349$. Since the smallest integer larger than $\sqrt{64349}$ is 254, we begin by letting $x = 254$. But since $254^2 - n = 167$ is not a perfect square, then 254 is not the correct value for x. Next, we try $x = 255$. Since $255^2 - n = 676 = 26^2$ is a perfect square, then the correct values of x and y are $x = 255$ and $y = 26$. Thus, the prime factors of n are $x + y = 281$ and $x - y = 229$.   ■

In comparison of the problems of primality testing and integer factorization, we should mention that factoring a known composite is in general significantly more time-consuming than finding a prime of approximately the same size. We have stated several times that the security of the RSA cryptosystem is based on the apparent difficulty of factoring a number that is the product of two very large distinct primes. To be more precise, the security of the RSA cryptosystem is based on the fact that it is apparently much more time-consuming for an intruder to factor the publicly known value of $n = pq$ than for the intended recipient of the message to choose p and q. (We use the word "apparently" because it has never been conclusively proven that factorization is significantly more time-consuming. Evidence, however, strongly suggests this.)

7.7 A Note on Digital Signatures

When the idea of public-key cryptography was developed, one way in which it was envisioned that it could be used was as follows. Suppose a group of people all wish to be able to communicate spontaneously with each other across a series of insecure lines of communication. For illustration, suppose they wish to use the RSA system to encipher their messages. To use the RSA system most effectively, each person in the group could choose their own secret primes p and q, form their own personal value for $n = pq$, and choose their own personal encryption exponent a. Each person in the group could then make their values of n and a public knowledge. Then, whenever a person in the group wanted to send another person in the group a secret message, they could use the intended recipient's public values of n and a to encipher the message. That way, only the intended recipient would be able to decipher the message. However, this leads to a problem for the intended recipient of the message, for the intended recipient would have no way to verify that the received message was sent by the person claiming to have sent it. This problem can be avoided as follows.

Suppose we wish to send the secret message P to a colleague across an insecure line of communication using RSA. Assume we have made public our personal RSA modulus n_1 and encryption exponent a_1 while keeping our decryption exponent b_1 secret, and our colleague has made public his or her personal RSA modulus n_2 and encryption exponent a_2 while keeping his or her decryption exponent b_2 secret. Suppose also that $n_1 < n_2$. To encipher our message, instead of applying our colleague's encryption exponent and modulus directly to the plaintext, we first apply our own decryption exponent and modulus. That is, instead of sending our colleague the ciphertext $P^{a_2} \bmod n_2$, we first compute $P_1 = P^{b_1} \bmod n_1$, and then send our colleague the ciphertext $C_1 = P_1^{a_2} \bmod n_2$. Our colleague can easily decipher this message by first applying his or her decryption exponent and modulus to obtain $P_1 = C_1^{b_2} \bmod n_2$, and then applying our publicly known encryption exponent and modulus to obtain $P = P_1^{a_1} \bmod n_1$. Since the decryption exponent b_1 we used in enciphering the message is known only to us, our colleague would know that only we could have enciphered the message. Because it has the effect of authenticating the message, applying our own decryption exponent and modulus in the encipherment of a message is sometimes called *signing* the message. For the case when $n_1 > n_2$, see Written Exercise 9.

Authentication of messages has been a very important and highly studied branch of cryptography for many years. In fact, it is interesting to note that in the title of their classic paper, "A Method for Obtaining Digital Signatures and Public-Key Cryptosystems", in which Rivest, Shamir, and Adleman introduced the RSA system, the notion of a digital signature was given precedence over that of a public-key cryptosystem.

7.8 The Diffie-Hellman Key Exchange

Recall that two people wishing to use the RSA cryptosystem to exchange a secret message across an insecure line of communication make their encryption exponent public knowledge. In this section we discuss a technique that can be used by two people to keep an RSA encryption exponent secret while communicating only across an insecure line of communication.[1]

There are several techniques by which two people can agree upon a cryptographic key secretly without having a secure way to communicate. One technique is the *Diffie-Hellman key exchange*, a process presented by W. Diffie and M. Hellman in their classic paper, "New Directions in Cryptography", in which they introduced the idea of public-key cryptography. In order to describe a way of incorporating this key exchange system with

---

[1] Copyright 1999 by COMAP, Inc. This material appeared in the spring 1999 issue of *UMAP* (see [10]).

the RSA system, suppose we wish to receive a secret message from a colleague using RSA. Furthermore, suppose we and our colleague would like to agree upon our encryption exponent secretly while communicating only across an insecure line of communication. We can accomplish this by the following steps in the Diffie-Hellman key exchange.

1. We choose primes p and q, form $n = pq$, and choose a positive integer $k < n$ with $(k, n) = 1$. We then send the values of k and n to our colleague across the insecure line of communication.

2. We choose a positive integer $r < n$, compute $k^r \bmod n$, and send the result to our colleague while keeping r secret. Meanwhile, our colleague chooses a positive integer $s < n$, computes $k^s \bmod n$, and sends the result to us while keeping s secret.

3. Both we and our colleague form the candidate encryption exponent $a = k^{rs} \bmod n$, which we compute as $(k^s)^r \bmod n$, and our colleague computes as $(k^r)^s \bmod n$. Since we know p and q, we can form $m = (p - 1)(q - 1)$ and determine if a is a valid RSA encryption exponent by determining if $(a, m) = 1$. If a is not a valid RSA encryption exponent, we repeat the process.

After we obtain a valid RSA encryption exponent, our colleague can then encipher his or her message using the usual RSA encipherment procedure with encryption exponent a and modulus n.

Example 7.6 Suppose we choose primes $p = 83$ and $q = 101$ so that $n = 8383$ and $m = 8200$. Suppose also that we choose $k = 256$, and send k and n to our colleague. We then choose $r = 91$, compute $256^{91} \bmod 8383 = 2908$, and send the result to our colleague while keeping r secret. Meanwhile, our colleague chooses $s = 4882$, computes $256^{4882} \bmod 8383 = 1754$, and sends the result to us while keeping s secret. Both we and our colleague then form the candidate encryption exponent $a = 6584$, which we compute as $1754^{91} \bmod 8383 = 6584$, and our colleague computes as $2908^{4882} \bmod 8383 = 6584$. However, because this value of a is not a valid RSA encryption exponent (6584 is not relatively prime to m), we would inform our colleague that we must repeat the process. For the second attempt, suppose we choose the same values for p, q, and k. We then choose $r = 17$, compute $256^{17} \bmod 8383 = 5835$, and send the result to our colleague. Meanwhile, our colleague chooses $s = 109$, computes $256^{109} \bmod 8383 = 1438$, and sends the result to us. Both we and our colleague then form the candidate encryption exponent $a = 3439$, which we compute as $1438^{17} \bmod 8383 = 3439$, and our colleague computes as

5835^{109} mod $8383 = 3439$. Since this value of a is a valid RSA encryption exponent, we would confirm to our colleague that he or she could proceed with the usual RSA encipherment procedure.   ■

Note that in this key exchange system, we must assume the values of k, n, k^r mod n, and k^s mod n are known to intruders since they were all transmitted across an insecure line of communication. Hence, in order for this key exchange system to be secure, it should be an essentially impossible problem for an intruder to determine k^{rs} mod n from the knowledge of k, n, k^r mod n, and k^s mod n. This problem is called the *Diffie-Hellman problem*. It has been conjectured that the only way to solve the Diffie-Hellman problem in general is to solve the discrete logarithm problem. We discuss this problem next.

Discrete logarithms are important to consider when studying the Diffie-Hellman key exchange because the solution to a particular discrete logarithm problem leads directly to the solution to a corresponding Diffie-Hellman problem. Suppose we intercept transmissions between our enemy as they perform a Diffie-Hellman key exchange. That is, using the variables defined previously, suppose we intercept values of k, n, k^r mod n, and k^s mod n. We now consider the problem of determining r from the knowledge of k, n, and k^r mod n. In this scenario, r is called a *discrete logarithm* of k^r mod n to the base k, and the problem of determining r from the knowledge of k, n, and k^r mod n is called the *discrete logarithm problem*. Note that if we could solve this general discrete logarithm problem, then the preceding general Diffie-Hellman problem would also be solved, for we could determine r from k^r mod n, and then compute $a = (k^s)^r$ mod n. However, solving the discrete logarithm problem is not an easy method for solving the Diffie-Hellman problem, for it can be argued that the best (fastest) way to solve the discrete logarithm problem with a composite modulus n involves first factoring n. Thus, the factorization problem that provides security to the RSA system also provides security to the Diffie-Hellman key exchange (as it has been presented in this section).

Many algorithms for computing discrete logarithms have been presented in literature. For small values of n, and some special large values of n (for example, powers of a small base), many mathematics software packages have pre-defined functions for directly computing discrete logarithms. The Maple function for computing discrete logarithms is the **mlog** function, which is part of the **numtheory** number theory package. If we enter the following commands in Maple

```
> with(numtheory):

> mlog(y, k, n);
```

for positive integers y, k, and n, Maple will return an integer r with the property that $y = k^r \bmod n$. (If no such integer exists, Maple will return *false*.) For example, the following command

> `mlog(1438, 256, 8383);`

$$109$$

indicates that $256^{109} \bmod 8383 = 1438$, a fact we used in Example 7.6.

Using **mlog** and provided n is small, an intruder who intercepts Diffie-Hellman key exchange transmissions could easily determine the resulting cryptographic key. For example, suppose an intruder intercepts the second set of transmissions $k = 256$, $n = 8383$, $k^r \bmod n = 5835$, and $k^s \bmod n = 1438$ from Example 7.6. The intruder could determine the resulting RSA encryption exponent a by using **mlog** as above to find that $s = 109$ satisfies $k^s \bmod n = 1438$, and then computing $a = 5835^{109} \bmod 8383 = 3439$.

In addition to the fact that **mlog** will in general run essentially forever for very large values of n (as will all known algorithms for computing discrete logarithms), there is another problem with using **mlog** to "undo" the operation of modular exponentiation. While it is true that entering the preceding general **mlog** command will cause Maple to return an integer r with the property that $y = k^r \bmod n$, this integer will not necessarily be the integer actually used in the modular exponentiation being undone. For example, in the first part of Example 7.6, we used the fact that $256^{4882} \bmod 8383 = 1754$. But the following command

> `mlog(1754, 256, 8383);`

$$782$$

indicates that also $256^{782} \bmod 8383 = 1754$. Hence, the number returned by Maple is not the exponent we used in the example. However, this would not pose a problem for an intruder who intercepts the first set of transmissions from Example 7.6, for the intruder could still find the resulting candidate encryption exponent a by computing $2908^{782} \bmod 8383 = 6584$. Thus, despite the fact that Maple returned an unexpected result, this result can still be used in the intruder's general procedure for determining the candidate encryption exponent. To see that this will be true in general, suppose an intruder uses the **mlog** command to try to find the exponent in a Diffie-Hellman key exchange transmission $k^s \bmod n$. Even if the number returned by Maple is $s' \neq s$, since it will be the case that $k^s \bmod n = k^{s'} \bmod n$, the intruder can still find the candidate encryption exponent a in the way illustrated above since it will also be the case that $a = (k^s)^r \bmod n = (k^{s'})^r \bmod n$.

Written Exercises

1. Consider the message, "ATTACK RIGHT FLANK".

   (a) Encipher this message using the RSA cryptosystem with primes $p = 11$ and $q = 23$ and encryption exponent $a = 7$. Use the correspondence α from Chapter 6 to convert the message into numerical form, and encipher each plaintext character separately as in Example 7.2.

   (b) Use the Euclidean algorithm to find the decryption exponent that corresponds to the encryption exponent in part (a).

   (c) Encipher this message using the RSA cryptosystem with primes $p = 83$ and $q = 131$ and encryption exponent $a = 3$. Use the correspondence α from Chapter 6 to convert the message into numerical form, and group the plaintext integers into blocks with four digits as in Example 7.3 before enciphering.

2. Suppose you wish to be able to receive messages from a colleague using the RSA cryptosystem. You begin the process by choosing primes $p = 17$ and $q = 29$ and encryption exponent $a = 153$. Verify that this value of a is a valid RSA encryption exponent, and use the Euclidean algorithm to find the corresponding decryption exponent.

3. Suppose your enemy is exchanging messages using the RSA cryptosystem, and you intercept their modulus $n = 33$, encryption exponent $a = 7$, and the following ciphertext: 27 8 20 29 16 16 9 13 20 13 0 8 30 16 13. Decipher this message. (The correspondence α from Chapter 6 was used to convert the message into numerical form, and each plaintext character was enciphered separately as in Example 7.2.)

4. Suppose you wish to be able to receive messages from a colleague using the RSA cryptosystem. You begin the process by choosing primes $p = 47$ and $q = 59$ and encryption exponent $a = 1779$. Suppose you also determine the corresponding decryption exponent $b = 3$, and you receive the following ciphertext from your colleague: 0792 2016 0709 0464 1497 1086 2366 0524. Decipher this message. (The correspondence α from Chapter 6 was used to convert the message into numerical form, and the plaintext integers were grouped into blocks with four digits as in Example 7.3 before being enciphered.)

5. Consider the modular exponentiation

$$1302182 0^{13783} \quad = \quad 91518013 \bmod 363794227$$

that enciphered the message in Example 7.4. Find the exact number of multiplications the technique for efficient modular exponentiation illustrated in Section 7.4 requires to perform this calculation.

6. Show that the technique for efficiently computing $P^a \bmod M$ illustrated in Section 7.4 requires in general at most $2 \log_2 a$ multiplications.

7. Show that 15 is a pseudoprime to the base 4 but not a pseudoprime to the base 3.

8. Use Fermat factorization to find the two prime factors of the integer $n = pq = 321179$.

9. Suppose you wish to send the secret message P to a colleague across an insecure line of communication using the RSA cryptosystem. Assume you have made public your personal RSA modulus n_1 and encryption exponent a_1 while keeping your decryption exponent b_1 secret, and your colleague has made public his or her personal RSA modulus n_2 and encryption exponent a_2 while keeping his or her decryption exponent b_2 secret. Suppose also that $n_1 > n_2$.

   (a) Explain how the method described in Section 7.7 for digitally signing your message could fail.

   (b) Devise a method similar to the one described in Section 7.7 for digitally signing your message that could not fail.

10. Using primes $p = 5$ and $q = 7$, act as both people in the Diffie-Hellman key exchange system and agree upon a valid RSA encryption exponent a. List the results from all trials of the key exchange process, including trials that do not result in a valid encryption exponent.

11. Show that the set U_n of units in Z_n forms a multiplicative group.

12. Prove Theorem 7.1.

13. Prove Corollary 7.3.

Maple Exercises

1. Consider the message, "GO PACK".

   (a) Encipher this message using the RSA cryptosystem with 3-digit primes p and q and a valid 2-digit encryption exponent a of your choice. Use the correspondence α from Chapter 6 to convert the message into numerical form, and group the plaintext integers into blocks with four digits as in Example 7.3 before enciphering.

   (b) Encipher this message using the RSA cryptosystem with 4-digit primes p and q and a valid 3-digit encryption exponent a of your choice. Use the correspondence α from Chapter 6 to convert the message into numerical form, and group the plaintext integers into blocks with six digits before enciphering.

   (c) Encipher this message using the RSA cryptosystem with 7-digit primes p and q and a valid 4-digit encryption exponent a of your choice. Use the correspondence α from Chapter 6 to convert the message into numerical form, and group the plaintext integers into a single block as in Example 7.4 before enciphering.

2. Suppose your enemy is exchanging messages using the RSA cryptosystem, and you intercept their modulus $n = 86722637$, encryption exponent $a = 679$, and the following ciphertext: 35747828 20827476 55134021 85009695. Decipher this message. (The correspondence α from Chapter 6 was used to convert the message into numerical form, and the plaintext integers were grouped into blocks with six digits before being enciphered.)

3. Set up a parameterization of the RSA cryptosystem using primes p and q with at least 30 digits each. Choose a valid encryption exponent a, and determine a corresponding decryption exponent b. Then use this parameterization of the RSA system to encipher and decipher the message, "CANCEL MISSION WAIT FOR NEW ORDERS". (Use the correspondence α from Chapter 6 to convert the message into numerical form.)

4. Using a Maple **for** or **while** loop, find the smallest base to which the number 3215031751 is not a pseudoprime.

5. Using primes $p = 503$ and $q = 751$, act as both people in the Diffie-Hellman key exchange system and agree upon a valid RSA encryption exponent a. List the results from all trials of the key exchange process, including trials that do not result in a valid encryption exponent. Also, show how an intruder could use Maple to find the value of a.

Chapter 8

Elliptic Curve Cryptography

Recall from Section 7.8 that the security of the Diffie-Hellman key exchange system is based on the difficulty of solving the discrete logarithm problem. In this chapter we discuss a public-key cryptosystem whose security is also based on the difficulty of solving the discrete logarithm problem. This system, named the *ElGamal* cryptosystem for T. ElGamal who first published the system in 1985, has formed an important area of recent cryptographic research due to how elliptic curves can naturally be incorporated into the system.

8.1 The ElGamal Cryptosystem

Before discussing elliptic curves and how they can naturally be incorporated into the ElGamal system, we first describe the system in general and give two simple examples of it. In order to describe the ElGamal system, suppose two people wish to exchange a secret message across an insecure line of communication. They can accomplish this by the following steps in the ElGamal cryptosystem:

1. As with the RSA cryptosystem, the intended recipient of the message initiates the process. The intended recipient chooses a finite abelian group G and an element $a \in G$, then chooses a positive integer n, computes $b = a^n$ in G, and makes the group G and the values of a and b public knowledge.

2. Using some public method of conversion, the originator of the message converts his or her message into an equivalent element or list of elements in G. Suppose the message converts to the element $w \in G$. The originator of the message then chooses a positive integer k, computes $y = a^k$ and $z = wb^k$ in G, and sends the values of y and z to the intended recipient across the insecure line of communication.

3. Because the intended recipient of the message knows n, the intended recipient can recover w by computing zy^{-n} in G since

$$zy^{-n} = wb^k(a^k)^{-n} = w(ba^{-n})^k = w(1)^k = w.$$

(Note: If $|a| = m$, then y^{-n} can be determined as y^{m-n}.)

Although the preceding steps are specific to the ElGamal cryptosystem, the system can appear in many different forms due to the various types of groups that can be used for G. This is precisely how we will incorporate elliptic curves into the system. We will show in Section 8.3 that elliptic curves over finite fields form abelian groups with a specially-defined operation. Of course, it is not necessary to use this type of group in the system. The ElGamal system is especially easy to implement if G is chosen to be a group like the multiplicative group Z_p^* for prime p.

Example 8.1 Suppose we wish to use the ElGamal cryptosystem to send the message, "NCSU", to a colleague across an insecure line of communication.

1. Our colleague begins the process by choosing $G = Z_p^*$ for prime $p = 100000007$. Next, our colleague chooses $a = 180989$ and $n = 5124541$, computes $b = a^n \bmod p = 10524524$, and sends the values of p, a, and b to us.

2. Suppose we use the correspondence α from Chapter 6 to convert our message into a single block numerical equivalent. That is, suppose we convert our message into the numerical equivalent $w = 13021820$. We then choose $k = 3638997$, compute $y = a^k \bmod p = 73133845$ and $z = wb^k \bmod p = 83973114$, and send the values of y and z to our colleague.

3. Our colleague can easily verify that the polynomial $x - 180989$ is primitive in $Z_p[x]$. Hence, the order of $a = 180989$ in Z_p^* is $p - 1$. Thus, our colleague can recover w by computing $zy^{-n} \bmod p = zy^{(p-1)-n} \bmod p = 13021820$.

■

Note that in Example 8.1 we would have to assume that the values of p, a, b, y, and z were all public knowledge since they were all transmitted across an insecure line of communication. And for the system to be secure, an intruder must not be able to determine $zy^{-n} \bmod p$. Hence, an intruder must not be able to determine the value of n from intercepted values of p, a, and $b = a^n \bmod p$. But this is precisely the statement of the discrete logarithm problem we discussed in Section 7.8 with a prime modulus. That is, the security of the ElGamal system in Example 8.1 is based on an intruder not being able to solve the discrete logarithm problem we discussed in Section 7.8 with a prime modulus. We mentioned in Section 7.8 that the discrete logarithm problem with a large composite modulus is in general very difficult to solve. This is true, of course, with a large prime modulus as well.

Discrete logarithms and the discrete logarithm problem can be defined much more generally than how we defined them in Section 7.8. More generally, for any element x in a finite group G and an element $y \in G$ that is a power of x, any integer r that satisfies $x^r = y$ is called a "discrete logarithm of y to the base x", and the problem of determining an integer r that satisfies $x^r = y$ is called the "discrete logarithm problem". As we mentioned in Section 7.8, many algorithms for computing discrete logarithms have been presented in literature. However, in groups with extremely large order, even the fastest known discrete logarithm algorithms are in general extremely time-consuming. For example, the fastest known discrete logarithm algorithms would take millions of years to compute discrete logarithms in groups with approximately 10^{200} elements.

As we mentioned above, the ElGamal cryptosystem can appear in many different forms due to the various types of groups that can be used for G. Recall that the group Z_p^* used in Example 8.1 is the group of nonzero elements in the finite field Z_p. We close this section with an example of the ElGamal system using the group of nonzero elements in a more general finite field.

Example 8.2 Suppose we wish to use the ElGamal cryptosystem to send a secret message to a colleague across an insecure line of communication.

1. Our colleague begins the process by choosing the primitive polynomial $p(x) = x^5 + 4x + 2 \in Z_5[x]$. Then for the finite field $F = Z_5[x]/(p(x))$ with $5^5 = 3125$ elements, our colleague lets G be the multiplicative group F^*. Next, our colleague chooses $a = x$ and $n = 1005$, computes $b = a^n = 2x^4 + 4x^3 + x^2 + 4x + 2$ in G, and sends $p(x)$, a, and b to us.

2. Using some public method of conversion, we convert our message into the field element $w = x^4 + x^3 + 3 \in G$. We then choose $k = 537$,

compute $y = a^k = 2x^4 + x^3 + 4x + 4$ and $z = wb^k = x^4 + 3x^3 + 2x^2 + 3$ in G, and send y and z to our colleague.

3. Since $p(x)$ is primitive and $a = x$, then $|a| = |F^*| = 3124$. Hence, our colleague can recover w by computing $zy^{-n} = zy^{3124-n} = x^4 + x^3 + 3$.

∎

Note that an intruder could break the ElGamal cryptosystem in Example 8.2 by solving the discrete logarithm problem in the group F^*. Specifically, an intruder could break the system by finding a discrete logarithm of $b = 2x^4 + 4x^3 + x^2 + 4x + 2 \in F^*$ to the base $a = x$.

8.2 The ElGamal Cryptosystem with Maple

In this section we show how Maple can be used to perform the computations in Examples 8.1 and 8.2.

In Example 8.1 our colleague began the process by choosing $G = Z_p^*$ with prime $p = 100000007$. The following command defines this value of p and shows that it is prime.

```
> p := nextprime(100000000);
```

$$p := 100000007$$

Next, our colleague chose the following values for a and n.

```
> a := 180989:
> n := 5124541:
```

Our colleague then formed $b = a^n \bmod p$. Recall that this computation can be done in an efficient way by using the Maple &^ command as follows.

```
> b := a &^ n mod p;
```

$$b := 10524524$$

Our colleague then sent the values of p, a, and b to us. We converted our message into the following numerical equivalent w and chose the following value for k.

```
> w := 13021820:
> k := 3638997:
```

Next, we computed $y = a^k \bmod p$ and $z = wb^k \bmod p$.

```
> y := a &^ k mod p;
```

$$y := 73133845$$

```
> z := w*(b &^ k) mod p;
```

$$z := 83973114$$

We then sent the values of y and z to our colleague. Recall that the polynomial $x - 180989$ is primitive in $Z_p[x]$, and thus the order of $a = 180989$ in Z_p^* is $p - 1$. The following command verifies this.

```
> Primitive(x-a) mod p;
```

true

Finally, our colleague recovered w by computing $zy^{(p-1)-n} \bmod p$.

```
> z*(y &^ (p-1-n)) mod p;
```

$$13021820$$

In Example 8.2 our colleague began the process by choosing the primitive polynomial $p(x) = x^5 + 4x + 2 \in Z_5[x]$. The following commands define $p(x)$ and show that it is primitive in $Z_5[x]$.

```
> p := x -> x^5 + 4*x + 2:
> Primitive(p(x)) mod 5;
```

true

For the finite field $F = Z_5[x]/(p(x))$ of order $5^5 = 3125$, our colleague then let $G = F^*$. Next, our colleague chose $a = x$ and the following value for n.

```
> a := x:
> n := 1005:
```

Our colleague then formed $b = a^n$ in G. Recall that this computation can be done by using the Maple **Powmod** command as follows.

```
> b := Powmod(a, n, p(x), x) mod 5;
```

$$b := 2x^4 + 4x^3 + x^2 + 4x + 2$$

Our colleague then sent $p(x)$, a, and b to us. We converted our message into the following field element w and chose the following value for k.

```
> w := x^4 + x^3 + 3:
> k := 537:
```

Next, we computed $y = a^k$ in G.

```
> y := Powmod(a, k, p(x), x) mod 5;
```

$$y := 2x^4 + x^3 + 4x + 4$$

We then computed $z = wb^k$ in G. To do this, we can enter the following
Powmod and **Rem** commands.

```
> bk := Powmod(b, k, p(x), x) mod 5;
```

$$bk := 3\,x^4 + 2\,x^2 + 3\,x + 1$$

```
> z := Rem(w*bk, p(x), x) mod 5;
```

$$z := x^4 + 3\,x^3 + 2\,x^2 + 3$$

We then sent y and z to our colleague. Recall that since $p(x)$ is primitive
and $a = x$, then $|a| = 3124$. Hence, our colleague recovered w by computing
$zy^{3124-n} \in G$. We can perform this computation by entering the following
Powmod and **Rem** commands.

```
> yn := Powmod(y, 3124-n, p(x), x) mod 5;
```

$$yn := x^4 + 4\,x^3 + 4\,x^2 + 4\,x + 2$$

```
> Rem(z*yn, p(x), x) mod 5;
```

$$x^4 + x^3 + 3$$

8.3   Elliptic Curves

Elliptic curves have figured prominently in several types of mathematical
problems. For example, the recent proof of Fermat's Last Theorem by An-
drew Wiles employed elliptic curves. Elliptic curves have also played an important role in integer factorization, primality testing, and, more recently,
public-key cryptography. The idea of using elliptic curves in public-key
cryptography was first proposed by N. Koblitz and V. Miller in 1985.

Let F be a field not of characteristic 2 or 3, and suppose $c, d \in F$
such that $x^3 + cx + d$ has no multiple roots or, equivalently, such that
$4c^3 + 27d^2 \neq 0$. Then the set of ordered pairs $(x, y) \in F \times F$ of solutions
to the equation

$$y^2 = x^3 + cx + d \tag{8.1}$$

together with a special element denoted by $\overline{O}$ and called the *point at infinity* is called an *elliptic curve*. The significance of the element $\overline{O}$ will be
described below. An elliptic curve, when endowed with a specially-defined
operation, forms an abelian group. This operation is initially best viewed
geometrically when applied to an elliptic curve over the reals. For example,
consider the following graph of the ordered pairs (x, y) of solutions to the
equation $y^2 = x^3 - 6x$ over the reals.

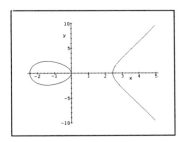

Note first that, as we would expect from the form of (8.1), this graph is symmetric about the x-axis. We now describe the operation that, when applied to the points on this graph and point at infinity $\overline{O}$, gives this elliptic curve E the structure of an abelian group. This operation is an addition operation and can be summarized as follows.

1. The point at infinity serves as the identity in the group. Thus, by definition, $P + \overline{O} = \overline{O} + P = P$ for all $P \in E$.

2. For any point $P = (x, y)$ on the graph of $y^2 = x^3 - 6x$, we define the negative of P to be $-P = (x, -y)$. This is illustrated in the following graph.

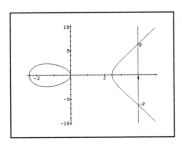

3. Suppose that P and Q are on the graph of $y^2 = x^3 - 6x$ with $P \neq \pm Q$, and that the line connecting P and Q is not tangent to the graph at P or Q. Then it is not difficult to show that the line connecting P and Q will intersect the graph at a unique third point R. We then define $P + Q = -R$. This is illustrated in the following graph.

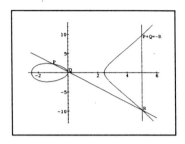

4. Suppose that P and Q are on the graph of $y^2 = x^3 - 6x$ with $P \neq \pm Q$, and that the line connecting P and Q is tangent to the graph at P. We then define $P + Q = -P$. This is illustrated in the following graph.

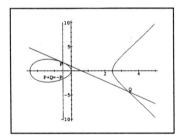

5. Suppose that P is on the graph of $y^2 = x^3 - 6x$ with $x \neq 0$, and that P is not a point of inflection for the graph. Then it is not difficult to show that the line tangent to the graph at P will intersect the graph at a unique second point R. We then define $P + P = -R$. This is illustrated in the following graph.

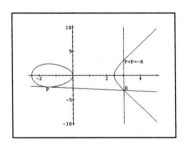

6. Suppose that P is on the graph of $y^2 = x^3 - 6x$, and that P is a point of inflection for the graph. We then define $P + P = -P$. This is illustrated in the following graph.

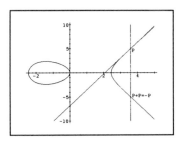

This operation is clearly commutative. The fact that this operation is associative is less obvious, and will be assumed.

Recall that for the ElGamal cryptosystem we need a finite abelian group. Hence, elliptic curves over the reals like the one illustrated above do not form groups that we could use in the ElGamal system. However, for an elliptic curve in which the underlying field F is finite, the elliptic curve will also be finite. For example, consider an elliptic curve in which the underlying field is Z_p for prime $p > 3$. Although the operation described above geometrically is applied specifically to an elliptic curve over the reals, this general operation gives any elliptic curve the structure of an abelian group. Of course, for an elliptic curve over Z_p, this operation cannot be described in the same way geometrically. However, the operation can be expressed algebraically.

Let p be a prime with $p > 3$, and suppose $c, d \in Z_p$ such that $x^3 + cx + d$ has no multiple roots or, equivalently, such that $4c^3 + 27d^2 \neq 0 \bmod p$. Let E be the elliptic curve of ordered pairs $(x, y) \in Z_p \times Z_p$ of solutions to (8.1) modulo p and point at infinity $\overline{O}$. It can be shown that the addition operation described above that gives E the structure of an abelian group can be expressed algebraically as follows. Recall first that $\overline{O}$ serves as the identity in the group. Now, let $P = (x_1, y_1)$ and $Q = (x_2, y_2)$ be elements in E. If $P = -Q$, then $P + Q = \overline{O}$. Otherwise, if $P = Q$, then we define $P + Q = (x_3, y_3)$ where

$$x_3 = \left(\frac{3x_1^2 + c}{2y_1} \right)^2 - 2x_1 \quad \bmod p, \tag{8.2}$$

$$y_3 = \left(\frac{3x_1^2 + c}{2y_1} \right)(x_1 - x_3) - y_1 \quad \bmod p. \tag{8.3}$$

And if $P \neq \pm Q$, then we define $P + Q = (x_3, y_3)$ where

$$x_3 = \left(\frac{y_2 - y_1}{x_2 - x_1} \right)^2 - x_1 - x_2 \quad \bmod p, \tag{8.4}$$

$$y_3 \quad = \quad \left(\frac{y_2 - y_1}{x_2 - x_1} \right) (x_1 - x_3) - y_1 \quad \text{mod } p. \tag{8.5}$$

For small primes p, we can construct elliptic curves over Z_p by trial and error. Let p be a prime with $p > 3$, and suppose $c, d \in Z_p$ such that $4c^3 + 27d^2 \neq 0 \bmod p$. We can then use the following steps to construct the solutions to (8.1) modulo p.

1. Determine which $x \in Z_p$ have the property that $z = x^3 + cx + d \bmod p$ is a perfect square in Z_p.

2. Find all $y \in Z_p$ such that $y^2 = z \bmod p$.

The values in Z_p^* that are perfect squares are called *quadratic residues*. Thus, the values of z determined in the preceding two steps are 0 and the quadratic residues in Z_p^*.

For the first preceding step, we consider the homomorphism $s(y) = y^2$ on Z_p^*. Note that the kernel of $s(y)$ is $K = \{x \mid x^2 = 1\} = \{1, -1\}$. Hence, $|K| = 2$, and the set $Q = \{z \in Z_p^* \mid z = s(y) \text{ for some } y \in Z_p^*\}$ of quadratic residues in Z_p^* has order $t = \frac{p-1}{2}$. Next, we consider the function $g(x) = x^t - 1$. If $z \in Q$, then $z = y^2 \bmod p$ for some $y \in Z_p$. Thus, $g(z) = z^t - 1 = y^{2t} - 1 = y^{p-1} - 1 = 0 \bmod p$ by Lagrange's Theorem. Hence, the t roots of $g(x)$ are precisely the t elements in Q. We summarize this test in the following lemma.

Lemma 8.1 *An element z is a quadratic residue in Z_p^* if and only if $z^{\frac{p-1}{2}} = 1 \bmod p$. Hence, z is a perfect square in Z_p if and only if $z = 0$ or $z^{\frac{p-1}{2}} = 1 \bmod p$.*

For the second preceding step, note that if $z = y^2 \bmod p$, it follows that $\left(z^{\frac{p+1}{4}} \right)^2 = y^{p+1} = y^2 = z \bmod p$. Therefore, for the second preceding step, if $p = 3 \bmod 4$, then we can find a square root of z by computing $z^{\frac{p+1}{4}} \bmod p$. We summarize this in the following lemma.

Lemma 8.2 *Suppose $p = 3 \bmod 4$. If z is a quadratic residue in Z_p^*, then $y = z^{\frac{p+1}{4}} \bmod p$ is a square root of z in Z_p^*. The only other square root of z in Z_p^* is $-y$.*

In summary, let p be a prime with $p > 3$ and $p = 3 \bmod 4$, and suppose $c, d \in Z_p$ such that $4c^3 + 27d^2 \neq 0 \bmod p$. Let E be the elliptic curve of ordered pairs (x, y) in $Z_p \times Z_p$ of solutions to (8.1) modulo p and point at infinity $\overline{O}$. Then for the set Q of quadratic residues in Z_p^*,

$$
\begin{aligned}
E \;=\; & \{(x, \pm y) \mid z = x^3 + cx + d \in Q \text{ and } y = z^{\frac{p+1}{4}} \bmod p\} \\
& \cup \{(x, 0) \mid x^3 + cx + d = 0\} \cup \{\overline{O}\}.
\end{aligned}
$$

Example 8.3 Let $p = 19$, and let E be the elliptic curve of ordered pairs $(x, y) \in Z_p \times Z_p$ of solutions to $y^2 = x^3 + x + 6$ modulo p and point at infinity $\overline{O}$. We can construct the ordered pairs in E as follows. First, by trial and error, we determine the values of x in Z_p for which $z = x^3 + x + 6 \bmod p$ is a quadratic residue in Z_p^*. For example, for $x = 0$, the value of z is $z = 0^3 + 0 + 6 \bmod p = 6$. Then since $z^{\frac{p-1}{2}} = 6^9 = 1 \bmod p$, Lemma 8.1 implies that $z = 6$ is a quadratic residue in Z_p^*. And for $x = 1$, the value of z is $z = 1^3 + 1 + 6 \bmod p = 8$. And since $z^{\frac{p-1}{2}} = 8^9 \neq 1 \bmod p$, Lemma 8.1 implies that $z = 8$ is not a quadratic residue in Z_p^*. By continuing this process, we can determine that the values of $x \in Z_p$ for which $z = x^3 + x + 6 \bmod p$ is a quadratic residue in Z_p^* are $x = 0, 2, 3, 4, 10, 12, 14$, and 18. Next, for each quadratic residue z in Z_p^* we must find the values of y in Z_p^* for which $y^2 = z \bmod p$. Since $p = 3 \bmod 4$, we can use Lemma 8.2 to do this. For example, for the quadratic residue $z = 6$ that results from $x = 0$, Lemma 8.2 implies that the square roots of z are $z^{\frac{p+1}{4}} \bmod p = 6^5 \bmod p = 5$ and -5. And since $z = 6$ results from $x = 0$, then the ordered pairs $(0, \pm 5)$ are in E. By repeating this process for each of the quadratic residues in Z_p^*, we can determine that the ordered pairs $(0, \pm 5)$, $(2, \pm 4)$, $(3, \pm 6)$, $(4, \pm 6)$, $(10, \pm 16)$, $(12, \pm 6)$, $(14, \pm 16)$, and $(18, \pm 17)$ are all in E. Also, by trial and error, we can determine that the only value of $x \in Z_p$ for which $z = x^3 + x + 6 = 0 \bmod p$ is $x = 6$. Hence, the only additional ordered pair in E is $(6, 0)$.

Now, suppose that we wish to compute the sum of the elements $P = (x_1, y_1) = (2, 4)$ and $Q = (x_2, y_2) = (10, 16)$ in E. Denote this sum by $P + Q = (x_3, y_3)$. Since $P \neq \pm Q$, then we can use (8.4) and (8.5) to find x_3 and y_3. We first compute $\dfrac{y_2 - y_1}{x_2 - x_1} \bmod p$ as follows. (Note: $8^{-1} = 12 \bmod p$ since $(8)(12) = 96 = 1 \bmod p$. This inverse can be found by using the Euclidean algorithm as illustrated in Section 7.1.)

$$
\frac{y_2 - y_1}{x_2 - x_1} = \frac{16 - 4}{10 - 2} = (12)(8)^{-1} = (12)(12) \bmod p = 11 \bmod p.
$$

Then, using (8.4) and (8.5), we have

$$
\begin{aligned}
x_3 &= 11^2 - 2 - 10 &&= 14 \bmod p, \\
y_3 &= 11(2 - 14) - 4 &&= 16 \bmod p.
\end{aligned}
$$

Hence, in this elliptic curve, $(2,4) + (10,16) = (14,16)$.

Next, for the element $P = (x_1, y_1) = (0,5)$ in E, suppose that we wish to compute the sum $P + P = (x_3, y_3)$. To do this, we can use (8.2) and (8.3). We first compute $\dfrac{3x_1^2 + c}{2y_1} \bmod p$ as follows. (Note: $10^{-1} = 2 \bmod p$ since $(10)(2) = 20 = 1 \bmod p$.)

$$
\frac{3x_1^2 + c}{2y_1} = \frac{3(0)^2 + 1}{(2)(5)} = (1)(10)^{-1} = (1)(2) \bmod p = 2 \bmod p.
$$

Then, using (8.2) and (8.3), we have

$$
\begin{aligned}
x_3 &= 2^2 - (2)(0) &&= 4 \bmod p, \\
y_3 &= 2(0 - 4) - 5 &&= 6 \bmod p.
\end{aligned}
$$

Hence, in this elliptic curve, $(0,5) + (0,5) = (4,6)$.　　　　■

Although elliptic curves over Z_p are not particularly easy to construct or even describe, their general structure is remarkably simple and specific. We summarize this general structure in the following theorem, which we state without proof.

Theorem 8.3 *Let E be an elliptic curve over Z_p for prime $p > 3$. Then E is isomorphic to the direct product $Z_{n_1} \times Z_{n_2}$ of the additive groups Z_{n_1} and Z_{n_2} for some integers n_1 and n_2 with $n_2 | n_1$ and $n_2 | (p - 1)$.*

As an example of Theorem 8.3, consider the elliptic curve E in Example 8.3. For this elliptic curve, $|E| = 18$, and thus the only possible values for n_1 and n_2 in Theorem 8.3 are $n_1 = 18$ and $n_2 = 1$, and $n_1 = 6$ and $n_2 = 3$. Since it can be verified that $(0,5) \in E$ generates all of the elements in E (as do several other elements in E), then E is cyclic. Hence, the correct values of n_1 and n_2 are $n_1 = 18$ and $n_2 = 1$, and E is isomorphic to the additive cyclic group Z_{18}.

Consider an elliptic curve E over Z_p for very large prime p. While constructing all of the elements in E is generally not possible, it is possible, although nontrivial, to compute the exact value of $|E|$ using a well-known algorithm by Schoof. Although Schoof's algorithm is beyond the scope of this book, we will mention a well-known result from Hasse that can be stated rather simply and yields upper and lower bounds on $|E|$. This result is commonly called *Hasse's Theorem*, which we state as follows without proof.

Theorem 8.4 *Let E be an elliptic curve over Z_p. Then*

$$p + 1 - 2\sqrt{p} \;\leq\; |E| \;\leq\; p + 1 + 2\sqrt{p}.$$

We close this section by mentioning one additional fact regarding elliptic curves. Recall that we began this discussion of elliptic curves by assuming that the underlying field F was not of characteristic 2 or 3, and that the cubic polynomial on the right-hand side of (8.1) had no multiple roots. Elliptic curves can also be defined over fields of characteristic 2 or 3; they are just not defined as the set of solutions to an equation of the exact form of (8.1). Specifically, if F is a field of characteristic 2, then an elliptic curve over F is defined as the set of ordered pairs $(x, y) \in F \times F$ of solutions to an equation of the form

$$y^2 + y \;=\; x^3 + cx + d \tag{8.6}$$

and point at infinity $\overline{O}$ where $c, d \in F$ and the cubic polynomial on the right-hand side of (8.6) is allowed to have multiple roots. And if F is a field of characteristic 3, then an elliptic curve over F is defined as the set of ordered pairs $(x, y) \in F \times F$ of solutions to an equation of the form

$$y^2 \;=\; x^3 + bx^2 + cx + d \tag{8.7}$$

and point at infinity $\overline{O}$ where $b, c, d \in F$ and the cubic polynomial on the right-hand side of (8.6) is not allowed to have multiple roots. Results analogous to those mentioned in this section also hold for elliptic curves over fields of characteristic 2 or 3.

8.4 Elliptic Curves with Maple

In this section we show how Maple can be used to construct the elliptic curve E in Example 8.3 and perform the elliptic curve addition operation.

We begin by defining the prime $p = 19$ and the values $c = 1$ and $d = 6$ for the elliptic curve equation (8.1).

```
> p := 19:
> c := 1:
> d := 6:
```

Recall that for the ordered pairs $(x, y) \in Z_p \times Z_p$ of solutions to (8.1) modulo a prime $p > 3$ and point at infinity $\overline{O}$ to form an elliptic curve, c and d must satisfy $4c^3 + 27d^2 \neq 0 \bmod p$. We verify this as follows.

```
> 4*c^3 + 27*d^2 mod p;
```

$$7$$

Next, we store the right-hand side of (8.1) as `eqn`.

```
> eqn := x^3 + c*x + d:
```

We now generate the elements in E that are ordered pairs $(x, y) \in Z_p \times Z_p$ of solutions to (8.1) modulo p. To generate these solutions we have provided the user-written procedure **epoints**, for which code is given in Appendix C.3. If this procedure is saved as the text file *epoints* in the directory from which we are running Maple, then we can include the **epoints** procedure in this Maple session by entering the following command.

```
> read epoints;
```

We can then generate the ordered pairs of solutions to (8.1) modulo p by entering the following command.

```
> ecurve := epoints(eqn, x, infinity, p);
```

$$ecurve := [0,5],[0,14],[2,4],[2,15],[3,6],[3,13],[4,6],[4,13],$$
$$[6,0],[10,16],[10,3],[12,6],[12,13],[14,16],[14,3],[18,17],$$
$$[18,2]$$

In the preceding command, the first parameter is the right-hand side of (8.1), and the second parameter is the variable used in the first parameter. The third parameter is a numerical value that indicates the number of solutions to (8.1) we wish the command to generate. If this parameter exceeds the total number of solutions to (8.1), then the command will generate all of the solutions to (8.1). By using **infinity** for this parameter, we guarantee that the command will generate all of the solutions to (8.1). The last parameter is the prime p.

Recall that the ordered pairs of solutions to (8.1) modulo p form all of the elements in E except the point at infinity $\overline{O}$. By entering the following command, we attach the representation 0 for the point at infinity to the list **ecurve** of elements in E.

```
> ecurve := ecurve, 0;
```

$$ecurve := [0,5],[0,14],[2,4],[2,15],[3,6],[3,13],[4,6],[4,13],$$
$$[6,0],[10,16],[10,3],[12,6],[12,13],[14,16],[14,3],[18,17],$$
$$[18,2],0$$

The following **nops** command returns the number of elements in E.

```
> nops([ecurve]);
```

$$18$$

To perform the elliptic curve addition operation defined in Section 8.3, we have provided the user-written procedure **addec**, for which code is given in Appendix C.3. Assuming this procedure is saved as the text file *addec* in the directory from which we are running Maple, then we can include the **addec** procedure in this Maple session as follows.

```
> read addec;
```

We can then compute the sum of the elements $[2, 4]$ and $[10, 16]$ in E by entering the following command.

```
> addec([2,4], [10,16], c, p);
```

$$[14, 16]$$

In the preceding command, the first two parameters are the elements in E we wish to add. The third and fourth parameters are the value of c from (8.1) and the prime p.

As another example of the **addec** procedure, in the next command we add the element $[0, 5]$ in E to itself.

```
> addec([0,5], [0,5], c, p);
```

$$[4, 6]$$

Next, we compute the sum of the element $[0, 5]$ and the point at infinity.

```
> addec([0,5], 0, c, p);
```

$$[0, 5]$$

And finally, we compute the sum of the elements $[0, 5]$ and $[0, 14]$ in E.

```
> addec([0,5], [0,14], c, p);
```

$$0$$

Note that the preceding output shows, as expected, that $[0, 5]$ and $[0, 14] = [0, -5]$ are inverses of each other in E.

We can verify that $[0, 5]$ is a cyclic generator for E as follows. We first assign the element $[0, 5]$ as the variable **gen**.

```
> gen := [0,5]:
```

We now construct the cyclic subgroup of E generated by **gen**. To do this, we first assign $[0, 5]$ also as the variable **temp** and store this element as the first entry in a table **cgroup**.

```
> temp := [0,5]:
> pct := 1:
> cgroup[pct] := temp:
```

Then by entering the following **while** loop, we construct the cyclic subgroup of E generated by **gen** and place these elements as the subsequent entries in **cgroup**. More specifically, by entering the following **while** loop, we compute multiples of **gen** using **addec** and place these multiples as the subsequent entries in **cgroup**. The loop terminates when the identity element 0 is obtained.

```
> while temp <> 0 do
>      temp := addec(temp, gen, c, p):
>      pct := pct + 1:
>      cgroup[pct] := temp:
> od:
> seq(cgroup[i], i = 1..pct);
```

$$[0,5],[4,6],[2,4],[3,6],[14,3],[12,13],[18,2],[10,3],[6,0],$$
$$[10,16],[18,17],[12,6],[14,16],[3,13],[2,15],[4,13],$$
$$[0,14],0$$

Since the preceding output is all of the elements in E, then $[0,5]$ is a cyclic generator for E.

8.5   Elliptic Curve Cryptography

If an elliptic curve over Z_p for some prime p is used as the group G in the ElGamal cryptosystem, the value of p would have to be extremely large in order for the system to be secure. More specifically, it is commonly accepted that G should contain a cyclic subgroup of order at least 2^{160} in order for the system to be secure. Constructing all of the elements in an elliptic curve over Z_p for extremely large p can be very time-consuming. However, to use an elliptic curve E over Z_p as the group G in the ElGamal system, it is not necessary to construct all of the elements in E. It is only necessary to find an element in E that has a relatively large order. Suppose we wish to use the ElGamal cryptosystem with an elliptic curve over Z_p as the group G in the system to send a secret message to a colleague across an insecure line of communication. Then the system could proceed as follows.

1. Our colleague begins the process by choosing a very large prime p and values for c and d that satisfy $4c^3 + 27d^2 \neq 0 \bmod p$. Let E be the elliptic curve of ordered pairs $(x, y) \in Z_p \times Z_p$ of solutions to

(8.1) modulo p and point at infinity $\overline{O}$. Our colleague then chooses an element $a \in E$ with relatively large order (which our colleague could verify by computing multiples of a in E). Next, our colleague chooses a positive integer n, computes $b = na = a + a + \cdots + a$ in E (note that we use the notation na for b instead of a^n since the elliptic curve operation is an addition operation), and sends the values of p, c, and d and the elements $a, b \in E$ to us.

2. Using some public method of conversion, we convert our message into an equivalent element $w \in E$. We then choose a positive integer k, compute $y = ka$ and $z = w + kb$ in E, and send the elements $y, z \in E$ to our colleague.

3. Our colleague can then recover w by computing $z - ny$ in E since

$$z - ny = w + kb - nka = w + kb - kb = w.$$

Example 8.4 Suppose we wish to use the ElGamal cryptosystem with an elliptic curve over Z_p as the group G in the system to send a secret message to a colleague across an insecure line of communication.

1. Our colleague begins the process by choosing $p = 19$ (for illustration, we use a very small value for p in this example), $c = 1$, and $d = 6$. Then the elliptic curve E of ordered pairs $(x, y) \in Z_p \times Z_p$ of solutions to (8.1) modulo p and point at infinity $\overline{O}$ is the elliptic curve in Example 8.3. Our colleague then chooses $a = (0, 5) \in E$, which, recall from Example 8.3, generates all of E. Next, our colleague chooses $n = 4$, computes $b = na = 4(0, 5) = (3, 6) \in E$ using the elliptic curve addition operation, and sends the values of p, c, and d and the elements $a, b \in E$ to us.

2. Using some public method of conversion, we convert our message into the element $w = (18, 17) \in E$. We then choose $k = 3$, compute $y = ka = 3(0, 5) = (2, 4)$ and $z = w + kb = (18, 17) + 3(3, 6) = (14, 3)$ in E using the elliptic curve addition operation, and send the elements $y, z \in E$ to our colleague.

3. Our colleague can then recover w by computing

$$
\begin{aligned}
z - ny &= (14, 3) - 4(2, 4) &&= (14, 3) - (12, 6) \\
&&&= (14, 3) + (12, 13) \\
&&&= (18, 17)
\end{aligned}
$$

using the elliptic curve addition operation. ∎

Note that to break the ElGamal cryptosystem in Example 8.4, an intruder would need to determine the value of n from the knowledge of a and $b = na$ in E. That is, to break the ElGamal cryptosystem in Example 8.4, an intruder would need to solve the discrete logarithm problem (expressed using the additive notation na for a^n) in E. Of course, because the elliptic curve in this example contains so few elements, an intruder could break the system very easily by trial and error. However, if an elliptic curve with an extremely large number of elements was used in the system, and the element a was chosen with a very large order, then it would be extremely difficult (time-wise) for an intruder to break the system.

There is a practical difficulty with using an elliptic curve E over Z_p as the group G in the ElGamal cryptosystem. Recall that if the system is implemented as described above, then the plaintext must be converted to one of the elements in E before being enciphered. This obviously limits flexibility in formatting plaintexts, and could possibly require the generation of many elements in E. We can avoid this difficulty by using a variation of the ElGamal system due to Menezes and Vanstone. Suppose as before that we wish to use the ElGamal cryptosystem with an elliptic curve over Z_p as the group G in the system to send a secret message to a colleague across an insecure line of communication. The steps in the Menezes-Vanstone variation of the ElGamal system can be stated as follows.

1. The first step is the same as in the usual ElGamal system. That is, our colleague chooses a very large prime p and values for c and d that satisfy $4c^3 + 27d^2 \neq 0 \bmod p$. Then for the elliptic curve E of ordered pairs $(x, y) \in Z_p \times Z_p$ of solutions to (8.1) modulo p and point at infinity $\overline{O}$, our colleague chooses an element $a \in E$ with large order and a positive integer n, computes $b = na$ in E, and sends p, c, d, a, and b to us.

2. We convert our message into an equivalent ordered pair of numbers $w = (w_1, w_2) \in Z_p^* \times Z_p^*$ (which does *not* need to be an element in E). We then choose a positive integer k, compute $y = ka$ and $kb = (c_1, c_2)$ in E, and encipher our message as the ordered pair $z = (z_1, z_2) \in Z_p^* \times Z_p^*$ by computing

$$z = (z_1, z_2) = (c_1 w_1 \bmod p, c_2 w_2 \bmod p).$$

We then send the ordered pairs y and z to our colleague.

3. Our colleague can first recover the ordered pair $kb = (c_1, c_2)$ by computing ny in E since $ny = nka = kna = kb$ in E. Our colleague can then recover the message $w = (w_1, w_2)$ by computing

$(c_1^{-1} z_1 \bmod p, c_2^{-1} z_2 \bmod p)$ since

$$(c_1^{-1} z_1 \bmod p, c_2^{-1} z_2 \bmod p) = (c_1^{-1} c_1 w_1 \bmod p, c_2^{-1} c_2 w_2 \bmod p)$$
$$= (w_1, w_2).$$

(Note: The multiplicative inverses c_1^{-1} and c_2^{-1} modulo p can be found in general by using the Euclidean algorithm as illustrated in Section 7.1.)

Example 8.5 Suppose we wish to use the Menezes-Vanstone variation of the ElGamal cryptosystem with an elliptic curve over Z_p as the group G in the system to send a secret message to a colleague across an insecure line of communication.

1. Our colleague chooses the values $p = 19$, $c = 1$, and $d = 6$ so that the elliptic curve E of ordered pairs $(x, y) \in Z_p \times Z_p$ of solutions to (8.1) modulo p and point at infinity $\overline{O}$ is the elliptic curve in Example 8.3. Our colleague then chooses $a = (0, 5) \in E$ and $n = 4$, computes $b = na = (3, 6) \in E$, and sends p, c, d, a, and b to us.

2. We convert our message into the ordered pair $w = (5, 13) \in Z_p \times Z_p$ (which, note from Example 8.3, is not an element in E). We then choose $k = 3$, compute $y = ka = (2, 4)$ and $kb = (12, 6)$ in E, and encipher our message by computing

$$z = ((12)(5) \bmod p, (6)(13) \bmod p) = (3, 2).$$

We then send y and z to our colleague.

3. Our colleague can first recover kb by computing $ny = (12, 6)$ in E. Our colleague can then recover w by computing

$$((12^{-1})(3) \bmod p, (6^{-1})(2) \bmod p)$$
$$= ((8)(3) \bmod p, (16)(2) \bmod p)$$
$$= (5, 13) .$$

(Note: $12^{-1} = 8 \bmod p$ since $(12)(8) = 96 = 1 \bmod p$, and $6^{-1} = 16 \bmod p$ since $(6)(16) = 96 = 1 \bmod p$.) ∎

Note that with the elliptic curve E in Example 8.3, the usual ElGamal system allows only $|E| = 18$ possible plaintexts, while the Menezes-Vanstone variation of the system allows $|Z_p^*|^2 = 324$ possible plaintexts.

8.6   Elliptic Curve Cryptography with Maple

In this section we show how Maple can be used to do the computations in an example of the Menezes-Vanstone variation of the ElGamal cryptosystem with an elliptic curve over Z_p as the group G in the system.

Recall that to use Lemma 8.2 in constructing an elliptic curve (which is employed in the user-written procedure **epoints** we used in Section 8.4 and will use again in this section), the prime p must satisfy $p = 3 \bmod 4$. We begin this section by entering the following procedure, which creates a Maple command for this session with the name **p3mod4**. This procedure is designed to quickly generate a large prime p with $p = 3 \bmod 4$.

```
> p3mod4 := proc(s)
>     local t;
>     t := nextprime(s);
>     while t mod 4 <> 3 do
>         t := nextprime(t);
>     od:
>     RETURN(t);
> end:
```

The Maple procedure defined by the preceding commands takes as its input an integer s and returns the smallest prime p larger than s that satisfies $p = 3 \bmod 4$. For example, the following command defines p as the smallest prime larger than 220532496293778805800 that satisfies $p = 3 \bmod 4$. We will use this prime in our example.

```
> p := p3mod4(220532496293778805800);
```

$$p := 220532496293778805891$$

For this value of p, let E be the elliptic curve of ordered pairs $(x, y) \in Z_p \times Z_p$ of solutions to (8.1) modulo p and point at infinity $\overline{O}$ with $c = 1$ and $d = 6$. In the following commands we define these values for c and d, and verify that they satisfy $4c^3 + 27d^2 \neq 0 \bmod p$.

```
> c := 1:
> d := 6:
> 4*c^3 + 27*d^2 mod p;
```

$$976$$

Next, we store the right-hand side of (8.1) as **eqn**.

```
> eqn := x^3 + c*x + d:
```

For the ordered pair a in the system, we will use the first solution to (8.1) generated by the user-written procedure **epoints** that was introduced in

Section 8.4. We define this element next.

```
> read epoints;
> a := epoints(eqn, x, 1, p);
```

$$a := [\,0, 56750407271085204502\,]$$

We could easily verify that this element a has relatively large order in E by repeatedly applying the user-written procedure **addec** that was introduced in Section 8.4 for adding elliptic curve elements. We will not include this verification here.

Next, we define the following value for n that we will use to construct the ordered pair $b = na$ in E.

```
> n := 91530873521338:
```

To expedite the process of adding a to itself n times using the elliptic curve addition operation, we have provided the user-written procedure **elgamal**, for which code is given in Appendix C.3. If this procedure is saved as the text file *elgamal* in the directory from which we are running Maple, then we can include the **elgamal** procedure in this Maple session by entering the following command. (Note: Because **elgamal** calls and uses the **addec** procedure that was introduced in Section 8.4, the **addec** procedure must be saved as the text file *addec* in the same directory as *elgamal*.)

```
> read elgamal;
```

We can then construct the ordered pair $b = na$ by entering the following command.

```
> b := elgamal(a, n, c, p);
```

$$[\,88936959893700554040, 106879392491870047319\,]$$

The parameters in this command are the ordered pair a, the multiple n of a we are computing, the value of c from (8.1), and the prime p.

Next, we define the following value for k that we will use to construct the ordered pairs $y = ka$ and kb in E.

```
> k := 431235145514:
```

We can then construct the ordered pairs $y = ka$ and kb in E as follows.

```
> y := elgamal(a, k, c, p);
```

$$[\,41921046194776811649, 52283417773968786897\,]$$

```
> kb := elgamal(b, k, c, p);
```

$$[\,88498850550708417382, 90428938891656008815\,]$$

We now use the ordered pair *kb* to encipher the message, "REN-DEZVOUS AT NOON". We first apply the correspondence α from Chapter 6 to convert this message into a list of two-digit integers. Using α, this message converts into the following list of integers: 17 04 13 03 04 25 21 14 20 18 00 19 13 14 14 13. Next, we group these integers into two blocks of equal length, and place these blocks as entries in the following ordered pair *w*.

```
> w := [1704130304252114, 2018001913141413]:
```

We can then encipher the message by entering the following command.

```
> z := [ kb[1]*w[1] mod p, kb[2]*w[2] mod p ];
```

$$z := [\,79041720375143250245, 25557336104884537057\,]$$

To decipher the message, we first recover the ordered pair *kb* by computing *ny* in *E* as follows.

```
> ny := elgamal(y, n, c, p);
```

$$[\,88498850550708417382, 90428938891656008815\,]$$

We can then decipher the message by entering the following command.

```
> [ (ny[1]^(-1)*z[1]) mod p, (ny[2]^(-1)*z[2]) mod p ];
```

$$[\,1704130304252114, 2018001913141413\,]$$

Written Exercises

1. Suppose you wish to use the ElGamal cryptosystem with a group of the form Z_p^* for some prime p as the group G in the system to send a secret message to a colleague across an insecure line of communication. Your colleague sends you the values $p = 31$, $a = 13$, and $b = 9$, and you convert your message into the numerical equivalent $w = 20$. Using the value $k = 6$, construct the values of y and z you would then send to your colleague.

2. Suppose you wish to receive a secret message across an insecure line of communication from a colleague using the ElGamal cryptosystem with a group of the form Z_p^* for some prime p as the group G in the system. You send your colleague the values $p = 13$, $a = 2$, and $b = 2^3 = 8 \bmod p$, and your colleague converts his or her message into a numerical equivalent w and returns to you the values $y = 5$ and $z = 2$. Decipher the message (recover w).

3. Suppose you wish to use the ElGamal cryptosystem with the group of nonzero elements in a finite field as the group G in the system to send a secret message to a colleague across an insecure line of communication. Your colleague sends you the primitive polynomial $p(x) = x^2 + x + 2 \in Z_5[x]$ and the polynomials $a = x$ and $b = 4x$ in G, and you convert your message into the element $w = 2x + 4 \in G$. Using the value $k = 6$, construct the polynomials y and z you would then send to your colleague.

4. Suppose you wish to receive a secret message across an insecure line of communication from a colleague using the ElGamal cryptosystem with the group of nonzero elements in a finite field as the group G in the system. You send your colleague the primitive polynomial $p(x) = x^2 + x + 2 \in Z_5[x]$ and the polynomials $a = x$ and $b = x^8 = 3x + 1$ in G, and your colleague converts his or her message into an element $w \in G$ and returns to you the polynomials $y = 2x$ and $z = 4x + 4$. Decipher the message (recover w).

5. Let E be the elliptic curve of ordered pairs $(x, y) \in Z_{11} \times Z_{11}$ of solutions to $y^2 = x^3 + x + 1$ modulo 11 and point at infinity $\overline{O}$.

   (a) Construct the elements in E.

   (b) Compute the sum $(3, 8) + (4, 6)$ in E.

   (c) Compute the sum $(1, 6) + (1, 6)$ in E.

   (d) Compute the sum $(1, 6) + (1, 5)$ in E.

6. Let E be the elliptic curve of ordered pairs $(x, y) \in Z_{23} \times Z_{23}$ of solutions to $y^2 = x^3 + x + 7$ modulo 23 and point at infinity $\overline{O}$.

   (a) Use Theorem 8.3 to show that E is cyclic. (Note: $|E| = 18$.)

   (b) Use Theorem 8.4 to find upper and lower bounds on $|E|$.

7. Let E be the elliptic curve of ordered pairs $(x, y) \in Z_{11} \times Z_{11}$ of solutions to $y^2 = x^2 + 2x$ modulo 11 and point at infinity $\overline{O}$.

   (a) Construct the elements of E.

   (b) Is E cyclic? State the structure of E given by Theorem 8.3.

8. Suppose you wish to use the usual ElGamal cryptosystem with an elliptic curve over Z_p for some prime p as the group G in the system to send a secret message to a colleague across an insecure line of communication. Your colleague sends you the elements $a = (8, 9)$ and $b = (1, 6)$ in the elliptic curve E in Written Exercise 5, and you convert your message into the element $w = (4, 6) \in E$. Using the value $k = 2$, construct the elements $y, z \in E$ you would then send to your colleague. (Hint: $7^{-1} = 8 \bmod 11$.)

9. Suppose you wish use the Menezes-Vanstone variation of the ElGamal cryptosystem with an elliptic curve over Z_p for some prime p as the group G in the system to send a secret message to a colleague across an insecure line of communication. Your colleague sends you the elements $a = (8, 9)$ and $b = (1, 6)$ in the elliptic curve E in Written Exercise 5, and you convert your message into the ordered pair $w = (5, 7)$. Using the value $k = 2$, construct the ordered pairs $y \in E$ and z you would then send to your colleague. (See hint at end of Written Exercise 8.)

10. Suppose you wish to receive a secret message across an insecure line of communication from a colleague using the Menezes-Vanstone variation of the ElGamal cryptosystem with an elliptic curve over Z_p for some prime p as the group G in the system. You send your colleague the elements $a = (4, 6)$ and $b = 2a = (6, 6)$ in the elliptic curve E in Written Exercise 5, and your colleague converts his or her message into an ordered pair w and returns to you the ordered pairs $y = (1, 6)$ and $z = (10, 10)$. Decipher the message (recover w). (Hint: $3^{-1} = 4 \bmod 11$, and $8^{-1} = 7 \bmod 11$.)

11. Recall that for the set of ordered pairs $(x, y) \in Z_p \times Z_p$ of solutions to (8.1) modulo a prime $p > 3$ and point at infinity $\overline{O}$ to be an elliptic curve, the values of c and d must satisfy $4c^3 + 27d^2 \neq 0 \bmod p$. To demonstrate the importance of this condition, use the elliptic curve

addition operation to add the ordered pairs $(0, 1)$ and $(14, 0)$ of solutions to the equation $y^2 = x^3 + x + 1$ modulo 31. Explain why your answer shows the importance of the condition $4c^3 + 27d^2 \neq 0 \bmod p$ in the definition of an elliptic curve over Z_p for prime $p > 3$.

Maple Exercises

1. Suppose you wish to use the ElGamal cryptosystem with a group of the form Z_p^* for some prime p as the group G in the system to send a secret message to a colleague across an insecure line of communication. Your colleague sends you the values $p = 10000000019$, $a = 132$, and $b = 240246247$, and you convert your message into the numerical equivalent $w = 2324123$. Using the value $k = 398824116$, construct the values of y and z you would then send to your colleague.

2. Suppose you wish to receive a secret message across an insecure line of communication from a colleague using the ElGamal cryptosystem with a group of the form Z_p^* for some prime p as the group G in the system. You send your colleague the values $p = 10000000019$, $a = 132$, and $b = a^n = 5803048419 \bmod p$ with $n = 121314333$, and your colleague converts his or her message into a numerical equivalent w and returns to you the values $y = 9054696956$ and $z = 7432712113$. Decipher the message (recover w).

3. Suppose you wish to use the ElGamal cryptosystem with the group of nonzero elements in a finite field as the group G in the system to send a secret message to a colleague across an insecure line of communication. Your colleague sends you the primitive polynomial $p(x) = 3x^7 + 4x + 1 \in Z_5[x]$ and the polynomials $a = x$ and $b = 3x^5 + x^4 + 2x^3 + 4x$ in G, and you convert your message into the element $w = 2x^6 + 4x^5 + x^2 + x + 1 \in G$. Using the value $k = 1851$, construct the polynomials y and z you would then send to your colleague.

4. Suppose you wish to receive a secret message across an insecure line of communication from a colleague using the ElGamal cryptosystem with the group of nonzero elements in a finite field as the group G in the system. You send your colleague the primitive polynomial $p(x) = 3x^7 + 4x + 1 \in Z_5[x]$ and the polynomials $a = x$ and $b = x^n = 2x^6 + 3x^5 + 1$ in G with $n = 51801$, and your colleague converts his or her message into an element $w \in G$ and returns to you the polynomials $y = x^6 + 4x^5 + 3x^4 + x^3 + x + 2$ and

$z = 2x^6 + 2x^5 + 3x^4 + 3x^3 + 2x^2 + 3x + 2$. Decipher the message (recover w).

5. Let E be the elliptic curve of ordered pairs $(x, y) \in Z_{59} \times Z_{59}$ of solutions to $y^2 = x^3 + 31x + 21$ modulo 59 and point at infinity $\overline{O}$.

   (a) Construct the elements in E.

   (b) Use Theorem 8.3 to show that E is cyclic. Then explain why every element in E except $\overline{O}$ will be a cyclic generator for E.

   (c) Compute the sum $(42, 3) + (54, 6)$ in E.

   (d) Compute the sum $(42, 3) + (42, 3)$ in E.

   (e) Compute the sum $(42, 3) + (42, 56)$ in E.

6. Set up a parameterization of the Menezes-Vanstone variation of the ElGamal cryptosystem using an elliptic curve over Z_p for some prime p with at least 25 digits as the group G in the system. Then use this parameterization of the ElGamal system to encipher and decipher the message, "TARGET HIT SEND NEW ORDERS". (Use the correspondence α from Chapter 6 to convert the message into numerical form.)

Chapter 9

Polya Theory

In this chapter we discuss some results for counting orbits when a group acts on a set. Because the most celebrated result we mention is the Polya Enumeration Theorem, we will refer to the theory we discuss in this chapter as *Polya* theory.

We begin by stating a very simple example of the type of problem we consider in this chapter. Suppose we wish to construct a necklace with four colored beads, and that each bead can be either blue or green. If we assume that the beads can be rotated around the necklace, and that the necklace can be flipped over and worn, then how many different necklaces can we construct? To answer this question, suppose we stretch the necklace into the shape of a square with one bead at each corner. The following figures show the set X of 16 possible arrangements for the beads.

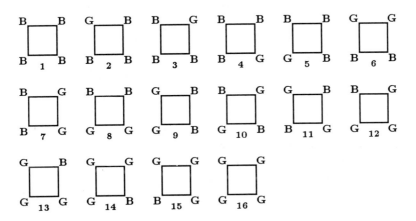

Of course, not all of these arrangements yield different necklaces. By rotating the beads around the necklace we can see that arrangements 2 through 5 are really the same necklace. Likewise, by flipping the necklace over we can see that arrangements 6 and 8 are really the same necklace. The movements of rotating the beads around the necklace and flipping the necklace over are called *rigid motions* of the necklace. We can also view these rigid motions as motions of the single figure

or, more specifically, motions of the set S of vertices of this figure. Note that each rigid motion of the necklace permutes the elements in X and S. Thus, we can represent these rigid motions by their permutations on X or S. We will use the permutations on S to answer questions like the one we posed at the start of this example. The advantage to using the permutations on S rather than X is that there are 16 elements in X, but only 4 elements in S. And this reduction in size would be much more important if the necklace contained more beads or if more colors were available for each bead. For example, if the necklace contained four beads but five colors were available for each bead, then there would be $5^4 = 625$ possible arrangements of the beads, but still only four vertices of the preceding general figure.

9.1   Group Actions

Recall that the set of rigid motions of a square forms a group with the operation of composition. In the following table we list the elements in this group G along with their permutations on S expressed as cycles. The rotations are counterclockwise. (Note that we include all cycles of length one. The significance of this will be apparent in Section 9.3.)

Element in G	Permutation on S
$\pi_1 = 90°$ rotation	(1234)
$\pi_2 = 180°$ rotation	$(13)(24)$
$\pi_3 = 270°$ rotation	(1432)
$\pi_4 =$ reflection across horizontal	$(12)(34)$
$\pi_5 =$ reflection across vertical	$(14)(23)$
$\pi_6 =$ reflection across 1–3 diagonal	$(24)(1)(3)$
$\pi_7 =$ reflection across 2–4 diagonal	$(13)(2)(4)$
$\pi_8 =$ identity	$(1)(2)(3)(4)$

Expressing the elements in G as permutations on X would in general require much longer notation. For example, the $90°$ rotation would be represented as a permutation on X by $(1)(2,5,4,3)(6,9,8,7)(10,11)(12,15,14,13)(16)$. Since it is true that each rigid motion $\pi \in G$ corresponds to unique permutations on S and X, in this chapter we will often write π when we mean to refer to one of the permutations. For example, by writing π_1, we could mean the $90°$ rotation, the permutation (1234) on S, or the permutation $(1)(2,5,4,3)(6,9,8,7)(10,11)(12,15,14,13)(16)$ on X. The context will make it clear which one we intend.

We now formalize our necklace example. Let S be a collection of objects, and let R be a set of elements called *colors* (not necessarily colors in the usual sense). A *coloring* of S by R is an assignment of a unique color to each element in S. That is, a coloring of S by R is a function $f : S \to R$. Note that if $|S| = n$ and $|R| = m$, then there will be m^n distinct colorings of S by R. We will denote by X the set of colorings of S by R. The set X of 16 possible arrangements of the beads in our necklace example is the set of 16 colorings of $S = \{$vertices of a square$\}$ by $R = \{$blue, green$\}$.

Now, consider a group G and a set Y. An *action* of G on Y is a mapping $Y \times G \to Y$ such that

1. $y(gh) = ((y)g)h$ for all $y \in Y$ and $g, h \in G$,

2. $(y)1 = y$ for all $y \in Y$, where 1 represents the identity in G.

In our necklace example, the group $G = \{$rigid motions of a square$\}$ acts on both $S = \{$vertices of a square$\}$ and $X = \{$colorings of S by $R\}$ where $R = \{$blue, green$\}$. As illustrated in this example, when a group G acts on a set Y, each element in G can be represented as a permutation on Y.

Lemma 9.1 *Suppose a group G acts on a set Y. For any $x, y \in Y$, define $x \sim y$ if there exists $g \in G$ for which $(x)g = y$. Then $\sim$ is an equivalence relation.*

Proof. Exercise.   ∎

As a consequence of Lemma 9.1, when a group G acts on a set Y, the set is decomposed into equivalence classes of elements that can be mapped to each other by elements in G. These equivalence classes are called *orbits*. When Y is a set of colorings, these orbits are also called *patterns*. The general type of problem we consider in this chapter can be viewed as counting the number of patterns when a group acts on a set of colorings.

In summary, suppose S is a set, R is a set of colors, and X is the set of colorings of S by R. When a group G acts on S, G also acts on X by $((x)f)\pi = ((x)\pi)f$ for all $x \in S$, $f \in X$, and $\pi \in G$. Two colorings $f, g \in X$ are equivalent if there exists $\pi \in G$ such that $((a)f)\pi = (a)g$ for all $a \in S$. Hence, two of the 16 colorings in our necklace example are equivalent if there is a rigid motion of a square that maps one to the other. To answer the question that began our necklace example, we must only count the number of patterns under this equivalence. From the list of colorings shown at the beginning of this chapter we can easily see that there are six such patterns: $\{1\}$, $\{2, 3, 4, 5\}$, $\{6, 7, 8, 9\}$, $\{10, 11\}$, $\{12, 13, 14, 15\}$, and $\{16\}$. With only 16 possible arrangements for the beads, it is not necessary to consider the group action of G on S or X to count the patterns. However, it would certainly not be practical to list all of the possible arrangements for the beads if the necklace had 10 beads and 12 colors were available for each bead. In this chapter we discuss how the idea of a group action can be used to count patterns without actually constucting the patterns.

9.2   Burnside's Theorem

Our goal in this chapter is to count the number of patterns when a group acts on a set of colorings. Counting the number of orbits when a group acts on a set is the focus of a fundamental result from Burnside. Before establishing this result, we first define some additional terms.

Suppose a group G acts on a set Y. Then for each element $\pi \in G$, we denote by $\mathrm{Fix}(\pi)$ the set of elements in Y that are fixed by π. That is, $\mathrm{Fix}(\pi) = \{y \in Y \mid (y)\pi = y\}$.

Example 9.1 Consider G acting on X in our necklace example. Using the notation π_i defined at the start of Section 9.1 for the elements in G, and the enumeration from the beginning of this chapter for the colorings in X, we list $\mathrm{Fix}(\pi_i)$ for each $\pi_i \in G$ in the following table.

$\pi_i \in G$	$\mathrm{Fix}(\pi_i)$	$|\mathrm{Fix}(\pi_i)|$
π_1	1, 16	2
π_2	1, 10, 11, 16	4
π_3	1, 16	2
π_4	1, 7, 9, 16	4
π_5	1, 6, 8, 16	4
π_6	1, 2, 4, 10, 11, 12, 14, 16	8
π_7	1, 3, 5, 10, 11, 13, 15, 16	8
π_8	X	16

■

Suppose again that a group G acts on a set Y. Then for each element $y \in Y$, we denote by $\text{Stab}(y)$ the subgroup of elements in G that fix y. That is, $\text{Stab}(y) = \{\pi \in G \mid (y)\pi = y\}$.

Example 9.2 Consider again G acting on X in our necklace example. We list $\text{Stab}(x)$ for each $x \in X$ in the following table.

$x \in X$	$\text{Stab}(x)$	$\|\text{Stab}(x)\|$
1	G	8
2	π_6, π_8	2
3	π_7, π_8	2
4	π_6, π_8	2
5	π_7, π_8	2
6	π_5, π_8	2
7	π_4, π_8	2
8	π_5, π_8	2
9	π_4, π_8	2
10	$\pi_2, \pi_6, \pi_7, \pi_8$	4
11	$\pi_2, \pi_6, \pi_7, \pi_8$	4
12	π_6, π_8	2
13	π_7, π_8	2
14	π_6, π_8	2
15	π_7, π_8	2
16	G	8

■

Note that the sum of the entries in the $|\text{Fix}(\pi_i)|$ column in Example 9.1 and sum of the entries in the $|\text{Stab}(x)|$ column in Example 9.2 are both 48. This equality is guaranteed in general by the following lemma.

Lemma 9.2 If a group G acts on Y, then $\sum_{\pi \in G} |\text{Fix}(\pi)| = \sum_{y \in Y} |\text{Stab}(y)|$.

Proof. Exercise. (Let $S = \{(y, \pi) \mid (y)\pi = y, \ y \in Y, \ \pi \in G\}$, and count $|S|$ in two ways – first by ranging through the possibilities for y, and then by ranging through the possibilities for π.) ■

Suppose again that G acts on Y. Then for $y \in Y$ we denote the orbit of y by $\text{Orb}(y)$. That is, $\text{Orb}(y) = \{x \in Y \mid x = (y)\pi \text{ for some } \pi \in G\}$.

Lemma 9.3 If a group G acts on Y, then $|G| = |\text{Stab}(y)| \cdot |\text{Orb}(y)|$ for each $y \in Y$.

Proof. Suppose g and h are in the same right coset of Stab(y). Hence, $g = \pi h$ for some $\pi \in$ Stab(y). Thus, $(y)g = (y)\pi h = (y\pi)h = (y)h$. On the other hand, suppose $(y)g = (y)h$ for some $g, h \in G$. Then $y = (y)hg^{-1}$, and $hg^{-1} \in$ Stab(y). Therefore, $hg^{-1} = \pi$ for some $\pi \in$ Stab(y). Hence, $h = \pi g$, and h and g are in the same right coset of Stab(y). In summary, g and h are in the same right coset of Stab(y) if and only if $(y)g = (y)h$. Thus, there is a bijection between the right cosets of Stab(y) and the elements in Orb(y). Using this and Lagrange's Theorem (Theorem 1.4), we conclude

$$
\begin{aligned}
|G| &= |\text{Stab}(y)| \cdot (\text{number of right cosets of Stab}(y)) \\
    &= |\text{Stab}(y)| \cdot |\text{Orb}(y)| \ .
\end{aligned}
$$

■

We now establish the following fundamental result for counting orbits when a group acts on a set. This result is due to Burnside, and thus we will call it *Burnside's Theorem*.

Theorem 9.4 *Suppose a group G acts on a set Y. Then the number of orbits in Y is* $\dfrac{1}{|G|} \displaystyle\sum_{\pi \in G} |Fix(\pi)|.$

Proof. Dividing both sides of the equation in Lemma 9.2 by $|G|$ yields

$$
\frac{1}{|G|} \sum_{\pi \in G} |\text{Fix}(\pi)| = \frac{1}{|G|} \sum_{y \in Y} |\text{Stab}(y)| \ .
$$

And by Lemma 9.3, we know that

$$
\frac{1}{|G|} \sum_{y \in Y} |\text{Stab}(y)| = \sum_{y \in Y} \frac{1}{|\text{Orb}(y)|} \ .
$$

Suppose there are s orbits in Y, which we denote by $O_1, O_2, \ldots, O_s$. Then if $x, y \in O_i$, it follows that $|\text{Orb}(x)| = |\text{Orb}(y)| = |O_i|$. But then

$$
\sum_{y \in O_i} \frac{1}{|\text{Orb}(y)|} = \frac{1}{|O_i|} + \cdots + \frac{1}{|O_i|} = 1.
$$

Thus, $\displaystyle\sum_{y \in Y} \frac{1}{|\text{Orb}(y)|} = s.$ ■

To see how Theorem 9.4 can be applied, consider G acting on X in our necklace example. From Example 9.1, we can see that $\displaystyle\sum_{\pi \in G} |\text{Fix}(\pi)| = 48.$

Then, since $|G| = 8$, it follows from Burnside's Theorem that the number of orbits in X is $48/8 = 6$. That is, as we have already seen, there are 6 distinct necklaces in our necklace example. While this result is certainly correct, in practice the result of Burnside's Theorem is not usually determined exactly in this manner. Specifically, in practice the value of $\sum_{\pi \in G} |\text{Fix}(\pi)|$ is not usually determined by actually constructing the sets $\text{Fix}(\pi)$ as we did in Example 9.1. To construct the table in Example 9.1 we referenced the list of all possible necklace arrangements shown at the beginning of this chapter. However, recall that in general we would like to be able to count orbits without having to list all of the possible arrangements. We discuss a method for doing this next.

9.3   The Cycle Index

In our necklace example, consider the rigid motion $\pi_7 = $ reflection across 2–4 diagonal. Note that if $x \in \text{Fix}(\pi_7)$, then x must have the same color bead at vertices 1 and 3, but can have any color bead at vertices 2 and 4. Hence, since two colors are available for the beads, then there will be $2 \cdot 2 \cdot 2 = 8$ colorings fixed by π_7. Thus, we can determine that $|\text{Fix}(\pi_7)| = 8$ without having to reference the list of all possible necklace arrangements shown at the beginning of this chapter. And we could determine $|\text{Fix}(\pi_7)|$ in the same way if more than two colors were available for the beads. For example, if five colors were available for the beads, then there would be $5 \cdot 5 \cdot 5 = 125$ colorings fixed by π_7. Note that $|\text{Fix}(\pi_7)|$ depends only on the number of colors available for the beads and the number of sets of vertices that can take arbitrary colors. Specifically, if a colors are available for the beads, then $|\text{Fix}(\pi_7)| = a^3$.

The preceding discussion can be generalized as follows. Suppose π is a rigid motion for which k sets of vertices can take arbitrary colors. Then if a colors are available, it follows that $|\text{Fix}(\pi)| = a^k$. It is easy to see the number of sets of vertices that can take arbitrary colors from the cycle representation of π as a permutation on the set S of vertices. For example, recall that in our necklace example the rigid motion π_7 can be represented as the permutation $(13)(2)(4)$ on S. Since there are three disjoint cycles in this representation for π_7, then there will be three factors of a in $|\text{Fix}(\pi_7)|$. In general, if there are k disjoint cycles in the representation of π as a permutation on S and a colors are available, then $|\text{Fix}(\pi)| = a^k$. This relates to the material in Section 9.2 because it provides us with a way to use Burnside's Theorem for counting patterns without having to refer to a list of all of the possible arrangements. Specifically, it states that the sum in the formula in Burnside's Theorem can be expressed as $\displaystyle\sum_{\pi \in G} |\mathrm{Fix}(\pi)| = \sum_{\pi \in G} a^{k_\pi}$,

where k_π is the number of disjoint cycles in the representation of π as a permutation on S.

Example 9.3 Consider G acting on S in our necklace example. From the table at the start of Section 9.1 we can see that the number of disjoint cycles in the representations of π_i as permutations on S are 1, 2, 1, 2, 2, 3, 3, and 4, respectively. Thus, with two colors available for the beads,

$$\sum_{\pi \in G} |\mathrm{Fix}(\pi)| = 2^1 + 2^2 + 2^1 + 2^2 + 2^2 + 2^3 + 2^3 + 2^4 = 48.$$

And if five colors were available for the beads instead of just two, then

$$\sum_{\pi \in G} |\mathrm{Fix}(\pi)| = 5^1 + 5^2 + 5^1 + 5^2 + 5^2 + 5^3 + 5^3 + 5^4 = 960.$$

Hence, by Burnside's Theorem there are $\displaystyle\frac{1}{|G|} \sum_{\pi \in G} |\mathrm{Fix}(\pi)| = \frac{1}{8} \cdot 960 = 120$

distinct necklaces if five colors are available for the beads. ∎

This process for computing $\displaystyle\sum_{\pi \in G} |\mathrm{Fix}(\pi)|$ can be refined as follows. Suppose $\pi \in G$ is a rigid motion that, when acting on S, is represented by the product of disjoint cycles of lengths $i_1, i_2, \ldots, i_t$. We then associate with π the monomial $f_\pi = x_{i_1} x_{i_2} \cdots x_{i_t}$. For example, with $\pi_7 = (13)(2)(4)$ in our necklace example, we associate the monomial $f_{\pi_7} = x_2 x_1 x_1 = (x_1)^2 x_2$. We then define the *cycle index* of G acting on S as

$$f(x_1, x_2, \ldots, x_w) = \frac{1}{|G|} \sum_{\pi \in G} f_\pi$$

where w is the length of the longest cycle in the representation of any $\pi \in G$ as a permutation on S. The cycle index is of interest to us because of the following theorem, which states how it can be used to count orbits when a group acts on a set.

Theorem 9.5 *Let S be a set, let R be a set of colors, and let X be the set of colorings of S by R. Suppose a group G acts on S with cycle index $f(x_1, x_2, \ldots, x_w)$. If $|R| = a$, then the number of patterns in X under the corresponding action of G on X is $f(a, a, \ldots, a)$.*

Proof. Exercise. ∎

Example 9.4 For our necklace example we list the monomial f_π for each $\pi \in G$ in the following table.

$\pi \in G$	f_π
(1234)	x_4
(13)(24)	$(x_2)^2$
(1432)	x_4
(12)(34)	$(x_2)^2$
(14)(23)	$(x_2)^2$
(24)(1)(3)	$(x_1)^2 x_2$
(13)(2)(4)	$(x_1)^2 x_2$
(1)(2)(3)(4)	$(x_1)^4$

Thus, the cycle index for this example is

$$f(x_1, x_2, x_3, x_4) \;=\; \frac{1}{8}\left(2x_4 + 3x_2^2 + 2x_1^2 x_2 + x_1^4\right).$$

Hence, if two colors are available for the beads, then there will be $f(2,2,2,2) = \frac{1}{8}(4 + 12 + 16 + 16) = 6$ distinct necklaces. And if five colors were available for the beads instead of just two, then there would be $f(5,5,5,5) = \frac{1}{8}(10 + 75 + 250 + 625) = 120$ distinct necklaces.   ■

Example 9.5 Suppose we wish to construct a necklace with six colored beads. As in the 4-bead necklace example, we assume that the beads can be rotated around the necklace, and that the necklace can be flipped over and worn. In this example, we use a cycle index to determine the number of distinct necklaces we can construct with a specified number of colors available for each bead. To do this, suppose we stretch the necklace into the shape of a hexagon with one bead at each corner. Consider the following general shape for the necklace.

Let G be the set of rigid motions of a hexagon, and let S be the set of vertices of the preceding general figure. In the following table, we list the elements $\pi \in G$, their cycle representations as permutations on S, and the associated monomials f_π.

$\pi \in G$	Permutation on S	f_π
60° rotation	(123456)	x_6
120° rotation	(135)(246)	$(x_3)^2$
180° rotation	(14)(25)(36)	$(x_2)^3$
240° rotation	(153)(264)	$(x_3)^2$
300° rotation	(165432)	x_6
reflection fixing 2,5	(13)(46)(2)(5)	$(x_1)^2(x_2)^2$
reflection fixing 1,4	(26)(35)(1)(4)	$(x_1)^2(x_2)^2$
reflection fixing 3,6	(15)(24)(3)(6)	$(x_1)^2(x_2)^2$
reflection across vertical	(16)(25)(34)	$(x_2)^3$
reflection across diagonal	(23)(14)(56)	$(x_2)^3$
reflection across diagonal	(12)(36)(45)	$(x_2)^3$
identity	(1)(2)(3)(4)(5)(6)	$(x_1)^6$

Thus, the cycle index for this example is

$$f(x_1, x_2, x_3, x_4, x_5, x_6) = \frac{1}{12}\left(2x_6 + 2x_3^2 + 4x_2^3 + 3x_1^2x_2^2 + x_1^6\right).$$

Hence, if two colors are available for the beads, then there will be $f(2, 2, 2, 2, 2, 2) = \frac{1}{12}(4 + 8 + 32 + 48 + 64) = 13$ distinct necklaces. And if five colors are available for the beads instead of just two, there will be $f(5, 5, 5, 5, 5, 5) = \frac{1}{12}(10 + 50 + 500 + 1875 + 15625) = 1505$ distinct necklaces. ∎

9.4   The Pattern Inventory

From Example 9.5, we can see that 1505 distinct necklaces can be constructed with six beads if five colors are available for each bead. Consider now the following question. How many of these 1505 distinct necklaces have beads with only three of the five possible colors? Or, more specifically, if the colors available for the beads are blue, green, red, white, and yellow, then how many of these 1505 distinct necklaces have exactly two red beads, three white beads, one yellow bead, and no blue or green beads? In this section, we discuss a way to answer such questions.

Let S be a set, let R be the set $\{C_1, C_2, \ldots, C_t\}$ of colors, and let X be the set of colorings of S by R. Suppose a group G acts on S with cycle index $f(x_1, x_2, \ldots, x_w)$. Then the simplified symbolic expression

$$f(C_1 + \cdots + C_t, \; C_1^2 + \cdots + C_t^2, \ldots\ldots, \; C_1^w + \cdots + C_t^w)$$

is called the *pattern inventory* of X. The pattern inventory of X allows us to answer questions like those posed at the start of this section. This is due to the following theorem, commonly called the *Polya Enumeration Theorem*.

Theorem 9.6 *Suppose the monomial* $k\, C_1^{i_1} C_2^{i_2} \cdots C_t^{i_t}$ *appears in the pattern inventory of* X. *Then there are* k *patterns in* X *in which* C_1 *appears* i_1 *times,* C_2 *appears* i_2 *times,* $\ldots$, *and* C_t *appears* i_t *times.*

Because the proof of the Polya Enumeration Theorem is extensive, before verifying this theorem we first show how it can be applied in our 4-bead necklace example.

Example 9.6 Consider our 4-bead necklace example with cycle index

$$f(x_1, x_2, x_3, x_4) = \frac{1}{8}\left(2x_4 + 3x_2^2 + 2x_1^2 x_2 + x_1^4\right).$$

Suppose that each bead can be either $B =$ blue or $G =$ green. Then the pattern inventory of the set X of colorings is

$$
\begin{aligned}
f(B + G&, B^2 + G^2, B^3 + G^3, B^4 + G^4) \\
&= \frac{1}{8}(2(B^4 + G^4) + 3(B^2 + G^2)^2 + 2(B + G)^2(B^2 + G^2) \\
&\qquad + (B + G)^4) \\
&= \cdots \\
&= B^4 + B^3 G + 2B^2 G^2 + BG^3 + G^4 .
\end{aligned}
$$

From this pattern inventory, we can easily see the number of distinct 4-bead necklaces that have prescribed numbers of blue and green beads. For example, because the term BG^3 appears in this pattern inventory with a coefficient of 1 and exponents of 1 on B and 3 on G, then there is only one distinct 4-bead necklace with one blue bead and three green beads. And because the term $2B^2 G^2$ appears in this pattern inventory with a coefficient of 2 and exponents of 2 on B and G, then there are two distinct 4-bead necklaces with two blue beads and two green beads.

Now, suppose that each bead can also be $R =$ red. Then the pattern inventory of the set of colorings is

$$f(B + G + R, B^2 + G^2 + R^2, B^3 + G^3 + R^3, B^4 + G^4 + R^4)$$

$$= \frac{1}{8}(2(B^4 + G^4 + R^4) + 3(B^2 + G^2 + R^2)^2 + \cdots + (B + G + R)^4)$$

$$= \cdots$$

$$= B^4 + B^3G + 2B^2G^2 + BG^3 + G^4 + B^3R + 2B^2R^2 + BR^3 + R^4$$
$$+ G^3R + 2G^2R^2 + GR^3 + 2BGR^2 + 2BG^2R + 2B^2GR \ .$$

For example, because the term $2BGR^2$ appears in this pattern inventory, then there are two distinct 4-bead necklaces with one blue bead, one green bead, and two red beads. Note that by adding the coefficients of all of the terms in this pattern inventory, we see that there are 21 distinct 4-bead necklaces if three colors are available for the beads. Also, note that by adding the coefficients of just the last ten terms in this pattern inventory, we see that 15 of these 21 distinct necklaces have at least one red bead. Finally, as we would expect, note that each term in the 2-color pattern inventory is present in the 3-color pattern inventory.   ∎

A pattern inventory can be used to answer both of the questions posed at the start of this section. Specifically, consider the 6-bead necklace example with cycle index

$$f(x_1, x_2, x_3, x_4, x_5, x_6) = \frac{1}{12}\left(2x_6 + 2x_3^2 + 4x_2^3 + 3x_1^2x_2^2 + x_1^6\right).$$

Suppose that each bead can be $B = $ blue, $G = $ green, $R = $ red, $W = $ white, or $Y = $ yellow. We showed in Example 9.5 that 1505 distinct necklaces can be constructed with six beads if five colors are available for each bead. Of these 1505 necklaces, the number that have two red beads, three white beads, one yellow bead, and no blue or green beads will be the coefficient of R^2W^3Y in the pattern inventory

$$f(B + G + R + W + Y, \ldots \ldots, B^6 + G^6 + R^6 + W^6 + Y^6).$$

Of course, it would not be easy to compute this pattern inventory by hand. However, with the help of a symbolic manipulator like Maple, this pattern inventory is very easy to compute. We show how Maple can be used to compute pattern inventories in Section 9.5.

We close this section with a discussion of why the Polya Enumeration Theorem is true. Rather than giving a formal proof of the theorem, which would be complicated and not intuitive, we give an informal discussion of why it is true with two colors. This discussion can be generalized in an obvious way for more than two colors.

Let S be a set of s vertices, let R be the set $\{B = \text{blue}, G = \text{green}\}$ of colors, and let X be the set of colorings of S by R. Suppose a group G of rigid motions acts on S with cycle index $f(x_1, \ldots, x_w)$. If $\pi \in G$ acts on S with a single cycle of length s, then for an element in X to be fixed by π, each of the vertices in S must be assigned the same color. We will keep a record of this by writing $B^s + G^s$, which we interpret as representing the fact that a coloring fixed by π must have either s blue vertices or s green vertices. For example, with $\pi_1 = (1234)$ in our 4-bead necklace example, we would write $B^4 + G^4$, which we interpret as representing the fact that a coloring fixed by π_1 must have either four blue beads or four green beads.

Now, suppose $\pi \in G$ acts on S with two cycles of lengths s_1 and s_2. Then for an element in X to be fixed by π, all of the s_1 vertices represented in the first cycle of π must be assigned the same color, and all of the s_2 vertices represented in the second cycle of π must also be assigned the same color. We will keep a record of this by writing

$$(B^{s_1} + G^{s_1})(B^{s_2} + G^{s_2}),$$

which we interpret as representing the fact that for a coloring to be fixed by π, the s_1 vertices represented in the first cycle of π must be all blue or all green (hence the first factor), while the s_2 vertices represented in the second cycle of π must also be all blue or all green (hence the second factor). Note that by expanding this expression we obtain the following.

$$(B^{s_1} + G^{s_1})(B^{s_2} + G^{s_2}) = B^{s_1+s_2} + B^{s_1}G^{s_2} + B^{s_2}G^{s_1} + G^{s_1+s_2}$$

The terms on the right-hand side of this equation represent the fact that for a coloring to be fixed by π, all of the vertices must be blue (hence the first term), or the s_1 vertices represented in the first cycle of π must be all blue and the s_2 vertices represented in the second cycle of π must be all green (hence the second term), or the s_1 vertices represented in the first cycle of π must be all green and the s_2 vertices represented in the second cycle of π must be all blue (hence the third term), or all of the vertices must be green (hence the fourth term). For example, with $\pi_2 = (13)(24)$ in our 4-bead necklace example, we would write

$$(B^2 + G^2)(B^2 + G^2) = B^4 + B^2G^2 + B^2G^2 + G^4.$$

The first and last terms on the right-hand side of this equation indicate that the colorings in our 4-bead necklace example with four blue beads and four green beads are fixed by π_2. The middle two terms indicate that there are also two colorings in our 4-bead necklace example with two blue beads and two green beads that are fixed by π_2.

More generally, suppose $\pi \in G$ acts on S with j cycles of lengths $s_1, s_2, \ldots, s_j$. Then for an element in X to be fixed by π, the vertices represented in each of these cycles must all be assigned the same color. We keep a record of this by writing

$$(B^{s_1} + G^{s_1})(B^{s_2} + G^{s_2}) \cdots (B^{s_j} + G^{s_j}),$$

whose factors we interpret as representing the fact that for a coloring to be fixed by π, the vertices represented in each of the cycles of π must be all blue or all green. Recall that for use in the cycle index we attach to π the monomial $f_\pi = x_{s_1} x_{s_2} \cdots x_{s_j}$. Note that the above expression can be viewed as $f_\pi(B + G, \ldots, B^t + G^t)$, where t is the length of the longest cycle in π. As we have demonstrated, each term in the expansion of the above expression represents a coloring fixed by π with the distribution of colors given by the bases to the number of vertices specified by the exponents. Hence, if we combine the terms in this expansion that are similar, the coefficient of each resulting term will be the total number of colorings fixed by π with the distribution of colors given by the bases to the number of vertices specified by the exponents. And if we sum over all $\pi \in G$ and combine similar terms, the coefficient of each resulting term will be the total number of colorings fixed by any $\pi \in G$ with the distribution of colors given by the bases to the number of vertices specified by the exponents.

We now claim that if the monomial $kB^{i_1}G^{i_2}$ appears in the pattern inventory of X, then there will be k patterns in X in which B appears i_1 times and G appears i_2 times. To see this, let Y be the subset of X that contains all of the colorings in which B appears i_1 times and G appears i_2 times. Since G clearly acts on Y, Burnside's Theorem states that

$$\text{Number of patterns in } Y = \frac{1}{|G|} \sum_{\pi \in G} |\text{Fix}(\pi)| \qquad (9.1)$$

with G acting on Y. It is precisely this number of patterns that we wish to determine. As we have just discussed, the coefficients in the expanded form of $f_\pi(B + G, \ldots, B^t + G^t)$ show the number of colorings fixed by π (i.e., $|\text{Fix}(\pi)|$) for any $\pi \in G$. Hence, if we add the $B^{i_1}G^{i_2}$ terms in $f_\pi(B + G, \ldots, B^t + G^t)$ for all $\pi \in G$, the coefficient of the result will be the sum in (9.1). If we then divide this coefficient by $|G|$, this will show the number of patterns in Y. More generally, we can find the number of patterns for any possible distribution of the colors B and G by adding and simplifying $f_\pi(B + G, \ldots, B^t + G^t)$ for all $\pi \in G$ and dividing by $|G|$. The coefficients of the result will show the number of patterns in X with the distribution of colors given by the bases of the terms to the number of vertices specified by the exponents of the terms. Finally, note that adding $f_\pi(B + G, \ldots, B^t + G^t)$ for all $\pi \in G$ and dividing the result by $|G|$ will yield exactly the cycle index $f(x_1, x_2, \ldots, x_w)$ of G acting on S evaluated at $B + G, \ldots, B^w + G^w$. Since this is how we defined the pattern inventory of X, the result is shown.

9.5 The Pattern Inventory with Maple

In this section, we show how Maple can be used to count patterns and construct pattern inventories. We consider the 6-bead necklace example with cycle index

$$f(x_1, x_2, x_3, x_4, x_5, x_6) = \frac{1}{12}\left(2x_6 + 2x_3^2 + 4x_2^3 + 3x_1^2 x_2^2 + x_1^6\right).$$

We begin by defining this cycle index. Note that we use brackets **[ ]** to obtain the appropriate subscripts.

```
> f := (1/12)*(2*x[6] + 2*x[3]^2 + 4*x[2]^3 + 3*x[1]^2*x[2]^2
> + x[1]^6);
```

$$f := \frac{1}{6}x_6 + \frac{1}{6}x_3^2 + \frac{1}{3}x_2^3 + \frac{1}{4}x_1^2 x_2^2 + \frac{1}{12}x_1^6$$

To convert this expression into a function that we can evaluate in the usual manner, we enter the following **unapply** command.[1]

```
> f := unapply(f, x[1], x[2], x[3], x[4], x[5], x[6]);
```

$$f := (x\_1, x\_2, x\_3, x\_4, x\_5, x\_6) \rightarrow$$
$$\frac{1}{6}x\_6 + \frac{1}{6}x\_3^2 + \frac{1}{3}x\_2^3 + \frac{1}{4}x\_1^2 x\_2^2 + \frac{1}{12}x\_1^6$$

Although the preceding command changes the variables in f from x_i to x_i, this has no effect on how f can be used. For example, we can find the number of distinct 6-bead necklaces if two colors are available for the beads by evaluating $f(2, 2, 2, 2, 2, 2)$ as follows.

```
> f(2, 2, 2, 2, 2, 2);
```

$$13$$

Hence, as we saw in Example 9.5, there are 13 distinct 6-bead necklaces if 2 colors are available for the beads. Suppose these colors are $B =$ blue and $G =$ green. To see how many of these 13 distinct necklaces have prescribed numbers of blue and green beads, we compute the following pattern inventory.

```
> simplify(f(B+G, B^2+G^2, B^3+G^3, B^4+G^4, B^5+G^5, B^6+G^6));
```

$$B^6 + G^6 + 3\,B^3 G^3 + 3\,B^4 G^2 + 3\,B^2 G^4 + B^5 G + B\,G^5$$

---

[1]The output displayed for this command was produced by Maple V Release 5. Previous releases of Maple yield output in which the variables are changed to $y1, y2, \ldots, y6$. This has no effect on how f can be used.

Thus, for example, because the term $3\,B^4\,G^2$ appears in this pattern inventory, then there are three distinct 6-bead necklaces with four blue beads and two green beads.

Now, suppose that the color $R = $ red is also available for the beads. We can find the number of distinct 6-bead necklaces if three colors are available for the beads by evaluating $f(3,3,3,3,3,3)$ as follows.

```
> f(3, 3, 3, 3, 3, 3);
```

$$92$$

Hence, there are 92 distinct 6-bead necklaces if 3 colors are available for the beads. To see how these colors are distributed in the patterns, we compute the following pattern inventory.

```
> simplify(f(B+G+R, B^2+G^2+R^2, B^3+G^3+R^3, B^4+G^4+R^4,
> B^5+G^5+R^5, B^6+G^6+R^6));
```

$$
\begin{aligned}
11\,B^2\,G^2\,R^2 &+ 6\,G\,R^3\,B^2 + 6\,G^3\,R\,B^2 + 3\,G\,R\,B^4 + 6\,B\,R^3\,G^2 \\
&+ 3\,B\,R\,G^4 + 6\,B^3\,R\,G^2 + 3\,B\,G\,R^4 + 6\,B\,G^3\,R^2 + 6\,B^3\,G\,R^2 \\
&+ B^6 + G^6 + 3\,B^3\,G^3 + 3\,B^4\,G^2 + 3\,B^2\,G^4 + B^5\,G + B\,G^5 \\
&+ 3\,B^3\,R^3 + 3\,G^3\,R^3 + 3\,B^4\,R^2 + 3\,B^2\,R^4 + 3\,G^4\,R^2 + 3\,G^2\,R^4 \\
&+ B^5\,R + B\,R^5 + G^5\,R + G\,R^5 + R^6
\end{aligned}
$$

Thus, for example, because the term $6\,G\,R^3\,B^2$ appears in this pattern inventory, then there are six distinct 6-bead necklaces with one green bead, three red beads, and two blue beads. Also, as we would expect, note that each term in the 2-color pattern inventory is present in the 3-color pattern inventory.

Finally, suppose that the color $W = $ white is also available for the beads. We can find the number of distinct 6-bead necklaces if four colors are available for the beads by evaluating $f(4,4,4,4,4,4)$ as follows.

```
> f(4, 4, 4, 4, 4, 4);
```

$$430$$

Hence, there are 430 distinct 6-bead necklaces if 4 colors are available for the beads. To see how these colors are distributed in the patterns, we compute the following pattern inventory.

```
> simplify(f(B+G+R+W, B^2+G^2+R^2+W^2, B^3+G^3+R^3+W^3,
> B^4+G^4+R^4+W^4, B^5+G^5+R^5+W^5, B^6+G^6+R^6+W^6));
```

$$
\begin{aligned}
11\,B^2\,G^2\,R^2 &+ 16\,B\,G\,R^2\,W^2 + 16\,B\,R\,G^2\,W^2 + 16\,B\,W\,G^2\,R^2 \\
&+ 16\,G\,R\,B^2\,W^2 + 16\,G\,W\,B^2\,R^2 + 16\,R\,W\,B^2\,G^2 \\
&+ 10\,B^3\,G\,R\,W + 10\,B\,G\,R\,W^3 + 10\,B\,G^3\,R\,W + 10\,B\,G\,R^3\,W \\
&+ 6\,G\,R^3\,B^2 + 6\,G^3\,R\,B^2 + 3\,G\,R\,B^4 + 6\,B\,R^3\,G^2 + 3\,B\,R\,G^4
\end{aligned}
$$

$$+ 6 B^3 R G^2 + 3 B G R^4 + 6 B G^3 R^2 + 6 B^3 G R^2 + B^6 + G^6$$
$$+ 3 B^3 G^3 + 3 B^4 G^2 + 3 B^2 G^4 + B^5 G + B G^5 + 3 B^3 R^3$$
$$+ 3 G^3 R^3 + 3 B^4 R^2 + 3 B^2 R^4 + 3 G^4 R^2 + 3 G^2 R^4 + B^5 R$$
$$+ B R^5 + G^5 R + G R^5 + R^6 + 3 B W G^4 + 6 B W^3 G^2$$
$$+ 3 B W R^4 + 6 G^3 W B^2 + 3 G W B^4 + 3 G R W^4 + 6 G R^3 W^2$$
$$+ 6 B^3 W G^2 + 6 G^3 R W^2 + 6 B W^3 R^2 + 6 B^3 W R^2$$
$$+ 3 B R W^4 + 6 B R^3 W^2 + 3 B^3 W^3 + 3 G^3 W^3 + 3 R^3 W^3$$
$$+ 3 B^4 W^2 + 3 B^2 W^4 + 3 G^4 W^2 + 3 G^2 W^4 + 3 R^4 W^2$$
$$+ 3 R^2 W^4 + W^6 + 6 B^3 R W^2 + 3 B G W^4 + 6 B G^3 W^2$$
$$+ 6 B^3 G W^2 + 11 G^2 R^2 W^2 + 11 B^2 R^2 W^2 + 11 B^2 G^2 W^2$$
$$+ 6 G W^3 B^2 + 6 G^3 W R^2 + 3 G W R^4 + 6 G W^3 R^2 + 3 R W B^4$$
$$+ 6 R^3 W B^2 + 6 R W^3 B^2 + 3 R W G^4 + 6 R^3 W G^2$$
$$+ 6 R W^3 G^2 + B^5 W + B W^5 + G^5 W + G W^5 + R^5 W + R W^5$$

Thus, for example, because the term $16 G R B^2 W^2$ appears in this pattern inventory, then there are 16 distinct 6-bead necklaces with 1 green bead, 1 red bead, 2 blue beads, and 2 white beads.

9.6 Switching Functions

In this section we show how the theory discussed in this chapter can be applied to the classification of switching functions. A *switching* function is a function $f : Z_2^n \to Z_2$. (More generally, a switching function is a process that can start with any number of inputs but has only two possible outputs. The preceding definition is sufficient for our purposes.) Switching theory was born in the first part of the 20th century due to the increasingly high volume of telephone calls being placed through local switchboards. The way switching functions were subsequently used by telephone companies led to their more recent use in the design of digital computers. It is in this area that switching functions as we have defined them (mapping from Z_2^n to Z_2) are most useful because of how computers store, send, and receive information as binary strings.

Although we will not describe any specific applications of switching functions, we will mention that in general it is desirable to keep a record of all possible switching functions. However, this poses a problem because, even for small values of n, switching functions are very numerous. Specifically, for each positive integer n, since $|Z_2^n| = 2^n$, there are 2^{2^n} switching functions. Hence, even for a value of n as small as 5, there are more than 4 billion switching functions. Because switching functions are so numerous, it would not be practical to keep a record of all of them. What is done instead is that an equivalence relation is defined on the set of switching functions. This breaks the set of switching functions into equivalence classes, and then a record can be kept of just one function from each equivalence class. In order to define the equivalence relation that is used in general, fix a positive integer n and let X be the set of colorings of Z_2^n by Z_2. Then X is the set of switching functions for the fixed value of n. Note that the symmetric group S_n acts on X by

$$\pi f(x_1, x_2, \ldots, x_n) = f(x_{\pi(1)}, x_{\pi(2)}, \ldots, x_{\pi(n)})$$

for $f \in X$ and $\pi \in S_n$ (see Written Exercise 11). Then for $f, g \in X$, we define $f \sim g$ if there exists $\pi \in S_n$ such that $\pi f = g$ (see Written Exercise 12).

Example 9.7 Let $n = 2$ so that $Z_2^n = Z_2^2 = \{00, 01, 10, 11\}$. Define $f : Z_2^2 \to Z_2$ by

$$\begin{aligned} f(0,0) &= 1 \\ f(0,1) &= 0 \\ f(1,0) &= 1 \\ f(1,1) &= 0. \end{aligned}$$

Let π be the cycle $(12) \in S_2$. Then $\pi f(x_1, x_2) = f(x_2, x_1)$. Hence,

$$\begin{aligned} \pi f(0,0) &= 1 \\ \pi f(0,1) &= 1 \\ \pi f(1,0) &= 0 \\ \pi f(1,1) &= 0. \end{aligned}$$

Thus, if we define $g : Z_2^2 \to Z_2$ by

$$\begin{aligned} g(0,0) &= 1 \\ g(0,1) &= 1 \\ g(1,0) &= 0 \\ g(1,1) &= 0, \end{aligned}$$

then $f \sim g$. ∎

Since switching functions are recorded in general by keeping a record of one function from each equivalence class, it is of obvious importance to know the number of equivalence classes for each value of n. This is precisely where the theory discussed in this chapter applies. To count the number of equivalence classes we can use a cycle index. And to count the number of equivalence classes in which the functions produce prescribed numbers of zeros and ones we can use a pattern inventory. We illustrate these ideas in the following example.

Example 9.8 In this example we consider the switching functions $f : Z_2^n \to Z_2$ with $n = 3$. We begin by noting that the elements in the symmetric group S_3 can be expressed as the following cycles.

$\pi \in S_3$	Cycle Representation
π_1	(1)(2)(3)
π_2	(12)(3)
π_3	(13)(2)
π_4	(1)(23)
π_5	(123)
π_6	(132)

Next, we apply each of the permutations in S_3 to the elements in the set $Z_2^3 = \{000, 001, 010, 011, 100, 101, 110, 111\}$. For example, applying π_2 to 011 yields 101 since π_2 flips the first and second entries and fixes the third. In the following table we list the results from applying each of the permutations in S_3 to the elements in Z_2^3. Also, in the first column of the following table we attach numerical labels to the elements in Z_2^3. We will use these labels to express the actions of the elements in S_3 on Z_2^3 as cycles.

Label	Z_2^3	π_1	π_2	π_3	π_4	π_5	π_6
1	000	000	000	000	000	000	000
2	001	001	001	100	010	100	010
3	010	010	100	010	001	001	100
4	011	011	101	110	011	101	110
5	100	100	010	001	100	010	001
6	101	101	011	101	110	110	011
7	110	110	110	011	101	011	101
8	111	111	111	111	111	111	111

Now, in the following table we list the actions of each of the permutations in S_3 on the labels of the elements in Z_2^3. For example, the action of π_2 is $(1)(2)(35)(46)(7)(8)$ since, using the labels in the preceding table, π_2 fixes elements 1, 2, 7, and 8; sends elements 3 and 5 to each other; and sends elements 4 and 6 to each other. Also, in the third column of the following table we list the monomials for the resulting cycle index.

$\pi \in S_3$	Action on Z_2^3	Monomial
π_1	$(1)(2)(3)(4)(5)(6)(7)(8)$	$(x_1)^8$
π_2	$(1)(2)(35)(46)(7)(8)$	$(x_1)^4(x_2)^2$
π_3	$(1)(25)(3)(47)(8)$	$(x_1)^4(x_2)^2$
π_4	$(1)(23)(4)(5)(67)(8)$	$(x_1)^4(x_2)^2$
π_5	$(1)(253)(467)(8)$	$(x_1)^2(x_3)^2$
π_6	$(1)(235)(476)(8)$	$(x_1)^2(x_3)^2$

Thus, the cycle index for this example is

$$f(x_1, x_2, x_3) \;=\; \frac{1}{6}\left(x_1^8 + 3x_1^4 x_2^2 + 2x_1^2 x_3^2\right).$$

Since there are two colors in this example (the numbers zero and one), then the total number of equivalence classes of switching functions with $n = 3$ is given by $f(2, 2, 2) = \frac{1}{6}(256 + 192 + 32) = 80$. To see how many of these equivalence classes contain functions that produce prescribed numbers of zeros and ones, we compute the following pattern inventory. Denote the colors by $A = $ zero and $B = $ one. Then the pattern inventory is

$$f(A + B, A^2 + B^2, A^3 + B^3)$$
$$= \frac{1}{6}((A + B)^8 + 3(A + B)^4(A^2 + B^2)^2 + 2(A + B)^2(A^3 + B^3)^2)$$
$$= \cdots$$
$$= A^8 + 4A^7B + 9A^6B^2 + 16A^5B^3 + 20A^4B^4 + 16A^3B^5 + 9A^2B^6$$
$$\quad + 4AB^7 + B^8.$$

Hence, for example, because the term $16A^5B^3$ appears in this pattern inventory, then there are 16 equivalence classes that contain functions that produce 5 zeros and 3 ones. ∎

9.7   Switching Functions with Maple

In this section we show how Maple can be used to count and classify equivalence classes of switching functions. We demonstrate using the results obtained in Example 9.8.

To construct the cycle index for a set of switching functions, we have provided the user-written procedure **switch**, for which code is given in Appendix C.4. Because **switch** calls and uses the user-written procedure **ppoly**, for which code is also given in Appendix C.4, both of the procedures

switch and **ppoly** must be saved as text files in the directory from which we are running Maple. If they are saved as the text files *switch* and *ppoly*, then we can include the **switch** procedure in this Maple session by entering the following command.

> read switch;

We can then construct the cycle index for the set of switching functions with $n = 3$ by entering the following command.[2]

> f := switch(3, x, 'maxsub');

$$f := \frac{1}{6} x_1{}^8 + \frac{1}{2} x_1{}^4 x_2{}^2 + \frac{1}{3} x_1{}^2 x_3{}^2$$

Note that the result is the cycle index in Example 9.8. In the preceding command, the first parameter specifies the value of n. The second parameter is the variable in whose terms the resulting cycle index is to be expressed. The third parameter is a variable defined by the command as the value of the largest subscript on the variables in the cycle index. (This would be important for larger values of n.)

Next, we convert the preceding cycle index into a function that we can evaluate in the usual manner. Note that in this command we include input parameters x[1] through x[3] because 3 is the largest subscript on the variables in the cycle index.)

> f := unapply(f, x[1], x[2], x[3]);

$$f := (x\_1, x\_2, x\_3) \rightarrow \frac{1}{6} x\_1{}^8 + \frac{1}{2} x\_1{}^4 x\_2{}^2 + \frac{1}{3} x\_1{}^2 x\_3{}^2$$

Since there are two colors, we can find the total number of equivalence classes of switching functions by evaluating $f(2, 2, 2)$ as follows.

> f(2, 2, 2);

$$80$$

To classify these equivalence classes, we compute the following pattern inventory. Denote the colors by $A = $ zero and $B = $ one. Then we can compute the pattern inventory by entering the following command.

> simplify(f(A+B, A^2+B^2, A^3+B^3));

$$4 B^7 A + 9 B^6 A^2 + 16 B^5 A^3 + 20 B^4 A^4 + 16 B^3 A^5 + 9 B^2 A^6$$
$$+ 4 B A^7 + A^8 + B^8$$

Note that the result is the pattern inventory in Example 9.8.

---

[2] Due to the tremendous number of switching functions for even moderately sized values of n, this routine, depending on your machine speed, can be very time-consuming for $n > 10$.

Written Exercises

1. Suppose you wish to construct a necklace with three colored beads, and each bead can be either red or white. Assume that the beads can be rotated around the necklace, and that the necklace can be flipped over and worn. Consider the following general shape for the necklace with one bead positioned at each corner.

Let G be the set of rigid motions of a triangle, let S be the set of vertices of the preceding general figure, and let X be the set of colorings of S by the colors $R = $ red and $W = $ white.

(a) For each $\pi \in G$, find Fix(π).

(b) For each $x \in X$, find Stab(x).

(c) Find the cycle index of G acting on S. Use this cycle index to determine the number of distinct necklaces you can construct.

(d) Find the pattern inventory of X.

(e) Suppose each bead can also be $B = $ blue. Determine the number of distinct necklaces you can construct with this additional color available. Also, find the new pattern inventory. According to this new pattern inventory, how many of the new distinct necklaces have at least one blue bead?

2. How many distinct necklaces can you construct with three beads if ten colors are available for each bead? Assume that the beads can be rotated around the necklace, and that the necklace can be flipped over and worn.

3. Suppose you wish to construct a necklace with five colored beads, and each bead can be either red or white. Assume that the beads can be rotated around the necklace, and that the necklace can be flipped over and worn.

(a) How many distinct necklaces can you construct?

(b) How many of the distinct necklaces in part (a) have two red beads and three white beads?

(c) How many of the distinct necklaces in part (a) have at least three white beads?

4. Suppose you wish to construct a building in the shape of a pentagon, and you will paint each side of the building one of ten different colors. Assume two buildings are equivalent if one could be rotated or flipped to look like the other. How many nonequivalent buildings can you construct?

5. Suppose you wish to construct a six-pointed star, and you will paint each point on the star either blue, green, or red. Assume two stars are equivalent if one could be rotated to look like the other. (Rotated only, not flipped!)

   (a) How many nonequivalent stars can you construct?

   (b) How many of the nonequivalent stars in part (a) have each of the three colors used on exactly two of the points?

   (c) How many of the nonequivalent stars in part (a) have each of the three colors used on at least one of the points?

6. Repeat Written Exercise 5 if you assume two stars are equivalent if one could be rotated or flipped to look like the other.

7. Find the number of equivalence classes of switching functions with $n = 4$.

8. Prove Lemma 9.1.

9. Prove Lemma 9.2.

10. Prove Theorem 9.5.

11. Let X be the set of switching functions for a fixed positive integer n. Show that the symmetric group S_n acts on X by

$$\pi f(x_1, x_2, \ldots, x_n) = f(x_{\pi(1)}, x_{\pi(2)}, \ldots, x_{\pi(n)})$$

   for $f \in X$ and $\pi \in S_n$.

12. Let X be the set of switching functions for a fixed positive integer n. For $f, g \in X$, define $f \sim g$ if there exists $\pi \in S_n$ such that $\pi f = g$ using the action of S_n on X defined in Written Exercise 11. Show that $\sim$ is an equivalence relation.

Maple Exercises

1. Suppose you wish to construct a necklace with five colored beads, and each bead can be either red, white, blue, or yellow. Assume that the beads can be rotated around the necklace, and that the necklace can be flipped over and worn. Let G be the set of rigid motions of a pentagon, let S be the set of vertices of a pentagon, and let X be the set of colorings of S by the colors R = red, W = white, B = blue, and Y = yellow. Find the pattern inventory of X. Then use this pattern inventory to determine the number of distinct necklaces you can construct with exactly two red beads, one white bead, one blue bead, and one yellow bead.

2. Find the number of equivalence classes of switching functions with $n = 5$. Also, determine how many of these equivalence classes contain functions that produce 30 zeros and 2 ones.

Appendix A

Basic Maple Tutorial

The purpose of this appendix is to introduce some basic commands, syntax, and programming concepts for Maple V Release 5. For a more thorough introduction to Maple, see [5] and [16].

A.1 Introduction to Maple

Maple is an advanced software tool designed for doing complicated mathematics quickly and precisely on a computer. To use Maple, you enter commands at "prompts" that can be identified as the following symbol that appears on the left in the body of a Maple worksheet.

>

When you access Maple, a Maple window will open that contains a prompt at which you can immediately begin performing mathematical operations. For example, you can use Maple to multiply the numbers 247 and 3756 by entering "247 * 3756;" as follows.

```
> 247 * 3756;
```
$$927732$$

When you enter a Maple command (assuming no syntax errors), Maple will perform the calculation and move the cursor to the next command line in the worksheet (which it will create if no subsequent command line exists). Each Maple command must end with either a semicolon or a colon. If you end a Maple command with a semicolon, Maple will display the result. If you use a colon, Maple will suppress the result. For example, if you enter the preceding command with a colon instead of a semicolon, Maple will respond as follows.

> 247 * 3756:

Despite the fact that no result is displayed after this command, the calculation was performed. Suppression of output is useful in many of the applications we discuss in this book.

If you do not include a semicolon or colon at the end of a Maple command, Maple will not perform the calculation and will respond with a warning message after a new prompt. For example, if you enter the preceding command without a semicolon or colon, Maple will respond with a warning message similar to the following.

> 247 * 3756
>
Warning, incomplete statement or missing semicolon

One of the great benefits of a computer package like Maple is that if you make a mistake in entering a command, you can go back to the command and correct the error. For example, you can remedy the preceding warning message by returning to the preceding command and entering it again with a semicolon or colon. If you enter the preceding command with a semicolon at the end of the command line, Maple will respond as follows.

> 247 * 3756;
>

927732

A.2   Arithmetic

Maple is an example of a *computer algebra* system. One feature of such a system is that it can be used as a very smart calculator. In particular, you can very easily use Maple to add, subtract, multiply, or divide numbers or algebraic expressions. These arithmetic operations can be performed in Maple by using the following symbols: + for addition, − for subtraction, * for multiplication, and / for division. Also, the operation of exponentiation can be performed in Maple by using the symbols ^ or **. As examples, two numbers or fractions can be added in Maple as follows.

> 253 + 7775;

8028

```
> 25/27 + 3/51;
```

$$\frac{452}{459}$$

Operations can be performed in Maple on the last entered result by using the percent symbol %.[1] For example, by entering the following command, we multiply the preceding result by 23.

```
> 23 * %;
```

$$\frac{10396}{459}$$

Two percent symbols listed together refer to the next-to-last entered result.

```
> 23 * %%;
```

$$\frac{10396}{459}$$

And we can evaluate 3^7 by entering either of the following commands.

```
> 3^7;
```

$$2187$$

```
> 3**7;
```

$$2187$$

Like other computer algebra systems, Maple uses exact arithmetic. Thus, if we divide two integers, Maple will return the exact answer as follows.

```
> 3235/7478;
```

$$\frac{3235}{7478}$$

The Maple **evalf** command can be used to obtain the decimal representation of a number. For example, the following command returns the decimal representation of the preceding number. The default number of digits displayed is ten.

```
> evalf(%);
```

$$.4326023001$$

---

[1] Maple V Release 5 is the first release of Maple that uses the percent symbol % to refer to the last entered result. Earlier releases of Maple use ditto marks " for this purpose.

To obtain more or fewer than ten digits, the desired number of digits must be specified. For example, the following command yields the first 20 digits in the decimal representation of the preceding fraction.

```
> evalf(%%, 20);
```

$$.43260230008023535705$$

As a final note in this section, we mention the fact that Maple will recognize only parentheses to enclose groups of objects with the basic arithmetic operations. Maple will *not* recognize square brackets or curly braces for this purpose. We illustrate this in the next two commands.

```
> 5^[15*(3+2)];
Error, non algebraic terms in power should be of the same type
```

```
> 5^(15*(3+2));
```

$$264697796016968855958850781462388113141059875488828125$$

A.3   Defining Variables and Functions

To assign a numerical value or expression to a variable in Maple, you must use the colon-equal := notation. For example, the following command assigns the value 5 to the variable y.

```
> y := 5;
```

$$y := 5$$

The variable y will then have this value throughout the current Maple session until y is assigned another value or its value is "unassigned". To display the contents of this variable, we must only enter the following command.

```
> y;
```

$$5$$

We can directly perform mathematical operations using the assigned variable y as illustrated in the next command.

```
> 4*y + 5;
```

$$25$$

The following command can be used to "unassign" the value of y. Note that we use back ticks ' in this command.

```
> y := 'y';
```

$$y := y$$

There are two ways to define functions in Maple. The most useful way is to use the minus-greater than -> notation. For example, the following command defines the function $f(x) = x^2$.

```
> f := x -> x^2;
```

$$f := x \rightarrow x^2$$

The reason this is the most useful way to define a function in Maple is because it allows standard functional notation like $f(5)$ to be used when evaluating the function at a particular value.

```
> f(5);
```

$$25$$

Functions can also be defined in Maple as expressions without using the -> notation. For example, the following command defines $f(x) = x^2$ as an expression.

```
> f := x^2;
```

$$f := x^2$$

The reason this method for defining a function in Maple is not as useful is because evaluating the function at a particular value then requires use of the Maple **subs** command. For example, to evaluate $f(5)$ we must enter the following command.

```
> subs(x=5, f);
```

$$25$$

For a function defined as an expression, standard functional notation is not understood by Maple and results in nonsense.

```
> f(x);
```

$$x(x)^2$$

```
> f(5);
```

$$x(5)^2$$

A.4   Algebra

Another benefit of Maple is that it allows entire algebraic expressions to be manipulated in the same way calculators manipulate numbers. Some important Maple commands for performing operations with algebraic expressions are **simplify** to simplify an expression, **expand** to expand an

expression, **factor** to factor an expression, and **solve** to solve an equation
or system of equations. Examples of these commands follow.

> simplify((x^3+1)/(x^2-x+1));

$$x + 1$$

> expand((x^2+1)*(x+1)*(x+3));

$$x^4 + 4\,x^3 + 4\,x^2 + 4\,x + 3$$

> factor(%);

$$(\,x^2 + 1\,)(\,x + 1\,)(\,x + 3\,)$$

> sol := solve(x^3-9*x^2+20*x=0 , x);

$$sol := 0, 4, 5$$

The output for the preceding **solve** command is returned by Maple as a
set whose elements can be chosen. For example, we can choose the second
element in **sol** by entering the following command.

> sol[2];

$$4$$

A.5   Case Sensitivity

Maple is case sensitive – it distinguishes between upper and lower case
characters in commands. For example, to factor the polynomial $x^2 - 2x - 3$,
we can enter the following command.

> factor(x^2-2*x-3);

$$(\,x + 1\,)(\,x - 3\,)$$

However, the next command does not yield the result.

> FACTOR(x^2-2*x-3);

$$\mathrm{FACTOR}(\,x^2 - 2\,x - 3\,)$$

Maple has several functions designed for doing modular arithmetic. For
many of these functions, the name of the function is the same as for the
nonmodular arithmetic function but with an upper case first letter. For
example, to factor the polynomial $x^2 - 2x - 3$ over the integers modulo 3,
we can enter the following command.

```
> Factor(x^2-2*x-3) mod 3;
```

$$x\,(\,x+1\,)$$

As another example, consider the Maple **irreduc** function, which returns *true* if a polynomial input is irreducible over the integers and *false* if not. The following command indicates, as expected, that $x^2 + 1$ is irreducible over the integers.

```
> irreduc(x^2+1);
```

$$true$$

However, the next command states that $x^2 + 1$ is not irreducible over the integers modulo 2.

```
> Irreduc(x^2+1) mod 2;
```

$$false$$

We can see how $x^2 + 1$ factors over the integers modulo 2 by entering the following **Factor** command.

```
> Factor(x^2+1) mod 2;
```

$$(\,x+1\,)^2$$

A.6 Help File

If you ever need to see information or an example regarding a particular Maple command, you can gain access to a help window for the command by entering the command name preceded by a question mark (and not followed by a semicolon). For example, the following command causes Maple to display a help window for the **factor** command.

```
> ? factor
```

A.7 Arrays and Loops

Arrays in Maple are data structures in which the elements are grouped sequentially. To create an array in Maple, we can use the **array** function. For example, the following command creates an array with four elements.

```
> a := array([5, 1, -4, 6]);
```

$$a := [\,5,\,1,\,-4,\,6\,]$$

Maple associates with each element in an array an integer that can be used
to access the element. To access an element in an array, we enter the name
of the array with the position of the element we wish to access in square
brackets. For example, the following command returns the third element
in the preceding array a.

> a[3];

$$-4$$

In this book we often create arrays that we use as vectors. In Maple,
the **vector** routine, which is part of the **linalg** linear algebra package (a
more detailed discussion of this package is given in Appendix B), can be
used to create vectors. Vectors function in Maple essentially the same way
as arrays, except that the integers Maple associates with the elements in a
vector always start with an integer index of 1, whereas arrays can have any
integer index. To illustrate the **vector** command, we first include the Maple
linalg package in this Maple session by entering the following command.

> with(linalg):

Then in the next command we create a vector with four elements.

> b := vector([1, 1, 3, -4]);

$$b := [\,1,\, 1,\, 3,\, -4\,]$$

The following command allows us to access the first element in b.

> b[1];

$$1$$

In the next command we create an empty array with storage for four elements.

> c := array(1..4);

$$c := \mathrm{array}(\,1..4, [\,]\,)$$

In Maple, loops are designed to repeat a specific command a specified
number of times. The most basic type of loop in Maple is a **for** loop. In
the following commands we enter a **for** loop in which we sequentially access
the elements in a and b, multiply each of the corresponding elements, and
store the results in c.

> for i from 1 to 4 do
>   c[i] := a[i]*b[i];
> od:

In this loop, the indexing element i starts at 1 and increases by 1 with
each passage through the loop. The loop terminates when i reaches the

upper index 4. Every **for** loop in Maple ends with a matching **od** statement (the reverse of the letters in the **do** statement). We use a colon after the **od** statement to prevent Maple from printing the intermediate calculations during the progression of the loop.

The Maple **evalm** command can be used to display the contents of a vector. For example, to display the vector c constructed in the preceding **for** loop, we enter the following command.

```
> evalm(c);
```

$$[5,\, 1,\, -12,\, -24]$$

Another useful type of loop in Maple is a **while** loop, which executes the commands inside the loop until a specified condition fails. For example, the following **while** loop constructs the same array c as the preceding **for** loop.

```
> i := 0:
> while i < 4 do
>     i := i + 1;
>     c[i] := a[i]*b[i];
> od:
```

This **while** loop executes the two commands inside the loop, each time incrementing i by 1, and terminates when i reaches 4. While loops also end with **od** statements. The next command shows that this **while** loop constructs the same array c as the preceding **for** loop.

```
> evalm(c);
```

$$[5,\, 1,\, -12,\, -24]$$

A.8 Conditional Statements

Conditional statements in Maple are designed to decide which provided commands to execute based on whether a provided statement is true or false. To demonstrate, we define the following numbers x and y.

```
> x := 13034021/29391911;
```

$$x := \frac{13034021}{29391911}$$

```
> y := 2483118283/4630112000;
```

$$y := \frac{2483118283}{4630112000}$$

The Maple **if** statement can be used to perform conditional statements. When combined with the **else** statement, it performs a double alternative step. For example, suppose we wish to determine which of x and y is larger. To do this, we can enter the following commands.

```
> if x > y then
>     print(x);
> else
>     print(y);
> fi:
```

$$\frac{2483118283}{4630112000}$$

The **if** statement in the preceding commands checks if x is greater than y. Since this is false, Maple executes the **print** command after the **else** statement. Maple **if** statements end with **fi** statements (the reverse of the letters in the **if** statement).

To produce multiple decision statements in Maple, we can use the **if** statement in combination with the **elif** and **else** statements. For example, the following Maple **for** loop compares the corresponding elements in the array **a** and vector **b** defined in Appendix A.7, and multiplies or divides the elements depending on which one is larger. Maple then stores the results in the array **c**. If the elements in **a** and **b** are equal, then the corresponding element in **c** is defined to be 0.

```
> for i from 1 to 4 do
>     if a[i] > b[i] then
>         c[i] := a[i]*b[i]:
>     elif a[i] < b[i] then
>         c[i] := a[i]/b[i]:
>     else
>         c[i] := 0:
>     fi:
> od:
```

We can see the resulting array **c** by entering the following command.

```
> evalm(c);
```

$$\left[5, \, 0, \, \frac{-4}{3}, \, -24\right]$$

A.9   Maple Procedures

A Maple procedure is a prearranged collection of commands that Maple executes together. To define a procedure in Maple we use the **proc** statement. In order to illustrate the syntax for writing a Maple procedure we construct the following procedure **dprod**, which is designed to compute the dot product of two vectors.

```
> dprod := proc(v1, v2)
>       local res, n, i;
>       res := 0:
>       n := linalg[vectdim](v1);
>       for i from 1 to n do
>             res := res + v1[i]*v2[i]:
>       od:
>       RETURN(res):
> end:
```

In this procedure, **v1** and **v2** are input vectors. Note that the **proc** statement is not ended with a semicolon. The **local** statement that appears at the start of this procedure defines variables whose values are used only in the procedure itself. After the procedure terminates, these variables will return to their assigned values, if any, from before the procedure was executed. The **RETURN** statement that appears at the end of this procedure specifies the value to be returned by the procedure to the calling program. If this statement is not included in the procedure, then the procedure will return its last computed result. To specify the end of the procedure, we use an **end** statement. And we use a colon after the **end** statement to prevent Maple from printing the commands and statements in the procedure after the procedure is entered or read in as text.

In the following commands, we define two vectors and demonstrate how the **dprod** procedure can be used to compute their dot product.

```
> vect1 := vector([1, 3, 5, 3, 6]);
```

$$vect1 := [\,1, 3, 5, 3, 6\,]$$

```
> vect2 := vector([7, 2, 1, 0, -1]);
```

$$vect2 := [\,7, 2, 1, 0, -1\,]$$

```
> dprod(vect1, vect2);
```

We mention one final note regarding how user-written procedures can be included in Maple sessions. First, the statements in any procedure can always be entered interactively, line by line, in a Maple session. However, if the procedure is very long (for example, see the procedures in Appendix C.4), this may not be practical. Another way to include a user-written procedure in a Maple session is by saving the text of the procedure as a text file, and reading the text file into a Maple session using a Maple **read** command. To do this in a UNIX-like environment, the procedure must be saved as a text file in the same directory in which the Maple program is running. For other operating systems, the proper location of the text file varies. Assuming we are working in a UNIX-like environment, if we have saved the text of the **dprod** procedure shown above as the text file *dprod*, then we can include the procedure in a Maple session by entering the following command.

```
> read(dprod);
```

The procedure can then be used as illustrated above.

Appendix B

Some Maple Linear Algebra Commands

Most of the Maple functions that deal with vector and matrix computations are part of the **linalg** linear algebra package. In this appendix, we give a brief introduction to some of these functions. We begin by entering the following command, which includes the **linalg** package in this Maple session.

```
> with(linalg):
```

By entering the preceding command, we gain access to all of the routines in the **linalg** package. Note that we used a colon at the end of this command. This suppresses the list of available linear algebra routines that would have been displayed had we used a semicolon. If you wish to see a list of the available routines, just enter the preceding command with a semicolon (or look at the help file on the **linalg** package – see Appendix A.6).

There are several ways to enter a matrix in Maple. In the following command, we use the Maple **matrix** function to define a 3×3 matrix A. The first two parameters in this command are the dimensions of the result. The remaining parameters enclosed in brackets are the entries in the matrix listed by consecutive rows.

```
> A := matrix(3, 3, [2, 5, 7, 3, 1, 7, 8, 1, 2]);
```

$$A := \begin{bmatrix} 2 & 5 & 7 \\ 3 & 1 & 7 \\ 8 & 1 & 2 \end{bmatrix}$$

Next, we again use the **matrix** function to define the matrix A, but this time we use slightly different syntax. In this command, we use double brackets within the **matrix** function. The outside brackets again contain the entries in the matrix listed by consecutive rows, but now each specific row is set off by another set of square brackets. This syntax does not require that we specify the dimensions of the result.

```
> A := matrix( [ [2, 5, 7], [3, 1, 7], [8, 1, 2] ] );
```

$$A := \begin{bmatrix} 2 & 5 & 7 \\ 3 & 1 & 7 \\ 8 & 1 & 2 \end{bmatrix}$$

In Maple, when a variable has been assigned as a number or an expression, it is usually possible to see this stored value or expression by entering the name of the variable. For example, note the following commands.

```
> a := 7;
```

$$a := 7$$

```
> a;
```

$$7$$

However, if a variable has been assigned as a matrix, then entering the name of the variable will not cause Maple to display the matrix. Instead, Maple will print only the name of the matrix. For example, if we try to view the matrix constructed above by entering the name of the variable A in which the matrix is stored, Maple will respond as follows.

```
> A;
```

$$A$$

To view the matrix A constructed above, we must use the Maple **evalm** function as follows.

```
> evalm(A);
```

$$\begin{bmatrix} 2 & 5 & 7 \\ 3 & 1 & 7 \\ 8 & 1 & 2 \end{bmatrix}$$

In the next command, we multiply each of the entries in A by the scalar 4. Again, Maple responds with only the name of the result.

```
> 4*A;
```

$$4\,A$$

But we can again use the Maple **evalm** function to see the result.

> evalm(4*A);

$$\begin{bmatrix} 8 & 20 & 28 \\ 12 & 4 & 28 \\ 32 & 4 & 8 \end{bmatrix}$$

Next, we enter the following 3×3 matrix B.

> B := matrix([[1, 0, 3], [-3, 5, 1], [2, 4, 1]]);

$$B := \begin{bmatrix} 1 & 0 & 3 \\ -3 & 5 & 1 \\ 2 & 4 & 1 \end{bmatrix}$$

To add matrices in Maple, we can use the usual + symbol.

> evalm(A + B);

$$\begin{bmatrix} 3 & 5 & 10 \\ 0 & 6 & 8 \\ 10 & 5 & 3 \end{bmatrix}$$

And to raise matrices to powers in Maple, we can use the usual ^ symbol.

> evalm(A^2);

$$\begin{bmatrix} 75 & 22 & 63 \\ 65 & 23 & 42 \\ 35 & 43 & 67 \end{bmatrix}$$

However, Maple distinguishes between scalar and matrix multiplication with different symbols. To multiply matrices in Maple, we must use the &* command rather than the usual * symbol.

> evalm(A &* B);

$$\begin{bmatrix} 1 & 53 & 18 \\ 14 & 33 & 17 \\ 9 & 13 & 27 \end{bmatrix}$$

The preceding operations of matrix addition, exponentiation, and multiplication can be combined when evaluating expressions in Maple. For example, for the matrices A and B constructed above, we can evaluate $A^3 - 4BA$ by entering the following command.

```
> evalm( A^3 - 4*B &* A );
```

$$\begin{bmatrix} 616 & 428 & 753 \\ 467 & 426 & 636 \\ 639 & 225 & 504 \end{bmatrix}$$

To enter vectors in Maple, we can use the **vector** function. For example, in the following command, we define a vector c of length 3 positions.

```
> c := vector([1, 4, 2]);
```

$$c := [1, 4, 2]$$

For the matrix A and vector c defined above, we can use the Maple **linsolve** command as follows to solve the equation $Ax = c$.

```
> x := linsolve(A, c);
```

$$x := \left[\frac{39}{205}, \frac{-144}{205}, \frac{121}{205} \right]$$

Also, the following command yields the inverse of A. Note that to obtain the inverse of A, we must only raise A to the power -1.

```
> invA := evalm(A^(-1));
```

$$invA := \begin{bmatrix} \dfrac{-1}{41} & \dfrac{-3}{205} & \dfrac{28}{205} \\ \dfrac{10}{41} & \dfrac{-52}{205} & \dfrac{7}{205} \\ \dfrac{-1}{41} & \dfrac{38}{205} & \dfrac{-13}{205} \end{bmatrix}$$

Since A is invertible, then we can also solve the equation $Ax = c$ by forming $x = A^{-1}c$ as follows.

```
> x := evalm(invA &* c);
```

$$x := \left[\frac{39}{205}, \frac{-144}{205}, \frac{121}{205} \right]$$

We close this appendix by mentioning how some special types of matrices can easily be defined in Maple. For example, in the following command, we define the 3×3 zero matrix.

```
> mat0 := matrix(3, 3, 0);
```

$$mat0 := \begin{bmatrix} 0 & 0 & 0 \\ 0 & 0 & 0 \\ 0 & 0 & 0 \end{bmatrix}$$

And in the following command, we define a 4×4 matrix containing all ones.

```
> mat1 := matrix(4, 4, 1);
```

$$mat1 := \begin{bmatrix} 1 & 1 & 1 & 1 \\ 1 & 1 & 1 & 1 \\ 1 & 1 & 1 & 1 \\ 1 & 1 & 1 & 1 \end{bmatrix}$$

Finally, in the following command, we use the Maple **diag** function to define the 5×5 identity matrix.

```
> id := diag(1, 1, 1, 1, 1);
```

$$id := \begin{bmatrix} 1 & 0 & 0 & 0 & 0 \\ 0 & 1 & 0 & 0 & 0 \\ 0 & 0 & 1 & 0 & 0 \\ 0 & 0 & 0 & 1 & 0 \\ 0 & 0 & 0 & 0 & 1 \end{bmatrix}$$

In general, the **diag** function yields a diagonal matrix with diagonal entries given in order by the parameters in the command.

Appendix C

User-Written Maple Procedures

C.1 Chapter 5 Procedures

```
rscoeff := proc(f, x, p, a)
    local g, i, j, ng, cg, fs, field, ftable;
    fs := 2^(degree(p));
    field := linalg[vector](fs);
    for i from 1 to fs-1 do
        field[i] := Powmod(a, i, p, a) mod 2:
    od:
    field[fs] := 0;
    ftable := table();
    for i from 1 to fs-1 do
        ftable[ field[i] ] := a^i:
    od:
    ftable[ field[fs] ] := 0;
    g := expand(f) mod 2;
    ng := 0;
    for j from 0 to degree(g,x) do
        cg := coeff(g, x, j):
        cg := ftable[ Rem(numer(cg), p, a) mod 2 ]
                / ftable[ Rem(denom(cg), p, a) mod 2 ];
        if degree(cg,a) < 0 then
```

```
            cg := cg * a^(fs-1);
        fi:
        if degree(cg,a) = (fs-1) then
            cg := cg/a^(fs-1);
        fi:
        ng := ng + cg*x^j:
    od:
    g := sort(ng mod 2, x);
    RETURN(g);
end:




binmess := proc(cw, n, p, a, ml)
    local i, j, bvect, vs, pco, dga, binmat, binvect;
    for i from 0 to ml do
        pco := coeff(cw, x, i):
        if pco <> 0 then
            dga := degree(pco, a):
            pco := Powmod(a, dga, p, a) mod 2:
        fi:
        vs := []:
        for j from 0 to n-1 do
            vs := [op(vs), coeff(pco, a, j)]:
        od:
        if i = 0 then
            binmat := linalg[matrix](1, n, vs):
        else
            binmat := linalg[stackmatrix](binmat, vs):
        fi:
    od:
    binvect := convert(binmat, vector);
    RETURN(evalm(binvect));
end:




bincoeff := proc(n, bmess)
    local i, j, k, bk, pcoeff, poly;
    pcoeff := []:
    bk := linalg[vectdim](bmess);
    i := 0;
    k := 0;
```

```
    while i < bk do
        poly := 0:
        for j from 1 to n do
            poly := poly + bmess[i+j]*a^(j-1):
        od:
        pcoeff := [op(pcoeff), poly]:
        k := k+1;
        i := k*n;
    od:
    RETURN(evalm(pcoeff)):
end:



rseuclid := proc(t, f, g, z, p, a)
    local q, r, rm1, rp1, um1, u, up1, vm1, v, vp1, i;
    rm1 := sort(Expand(f) mod 2);
    r := sort(Expand(g) mod 2);
    um1 := 1;
    u := 0;
    vm1 := 0;
    v := 1;
    read(rscoeff);
    while degree(r,z) >= t do
        rp1 := Rem(rm1, r, z, 'q') mod 2;
        rp1 := rscoeff(rp1, z, p, a);
        q := rscoeff(q, z, p, a);
        vp1 := expand(vm1 - v*q) mod 2;
        vm1 := v;
        v := sort(vp1, z);
        v := rscoeff(v, z, p, a);
        up1 := expand(um1 - u*q) mod 2;
        um1 := u;
        u := sort(up1);
        u := rscoeff(u, z, p, a);
        rm1 := r;
        r := sort(rp1, z);
        print('Q = ', q, '   R = ', r,
              '   V = ', v, '   U = ', u);
    od;
    print();
    RETURN(q, r, v, u):
end:
```

C.2   Chapter 7 Procedures

Note: The following two procedures are variations of procedures found in
the examples folder of the Maple V Release 3 student version (see [27])
produced by Waterloo Maple Inc. and the University of Waterloo.

```
to_number := proc(mess)
    local sl, cn, sn, ii, ntable;
    ntable := table(['a'=0, 'b'=1, 'c'=2, 'd'=3, 'e'=4,
                'f'=5, 'g'=6, 'h'=7, 'i'=8, 'j'=9, 'k'=10,
                'l'=11, 'm'=12, 'n'=13, 'o'=14, 'p'=15,
                'q'=16, 'r'=17, 's'=18, 't'=19, 'u'=20,
                'v'=21, 'w'=22, 'x'=23, 'y'=24, 'z'=25]):
    sl := length(mess);
    cn := 0;
    for ii from 1 to sl do
        sn := ntable[substring(mess, ii..ii)]:
        cn := 100*cn + sn:
    od:
    RETURN(cn):
end:


to_letter := proc(num)
    local cs, cn, sl, a, b, c, d, e, f, g, h, i, j, k,
          l, m, n, o, p, q, r, s, t, u, v, w, x, y, z,
          ltable, ans;
    ltable := table([0=a, 1=b, 2=c, 3=d, 4=e, 5=f, 6=g,
                7=h, 8=i, 9=j, 10=k, 11=l, 12=m,
                13=n, 14=o, 15=p, 16=q, 17=r, 18=s,
                19=t, 20=u, 21=v, 22=w, 23=x, 24=y,
                25=z]);
    cn := num;
    sl := floor(trunc(evalf(log10(cn)))/2) + 1:
    ans := '';
    for i from 1 to sl do
        cn := cn/100;
        cs := ltable[frac(cn)*100];
        ans := cat(cs, ans);
        cn := trunc(cn);
    od:
    RETURN(ans);
end:
```

C.3   Chapter 8 Procedures

```
epoints := proc(ec, x, ub, p)
    local ecurve, z, pct, k, i;
    pct := 0;
    for k from 0 to p-1 while pct <= ub do
        z := subs(x=k, ec) mod p;
        if z = 0 then
            pct := pct + 1;
            ecurve[pct] := [k, z];
        fi:
        if z &^ ((p-1)/2) mod p = 1 then
            z := z &^ ((p+1)/4) mod p;
            ecurve[pct+1] := [k, z];
            ecurve[pct+2] := [k, -z mod p];
            pct := pct + 2;
        fi:
    od:
    if pct > ub then
        pct := ub:
    fi:
    seq(ecurve[i], i = 1..pct):
end:
```

```
addec := proc(le, re, c, p)
    local i, cle, cre, lambda, res, x3, y3;
    cle := le mod p;
    cre := re mod p;
    if cle = 0 or cre = 0 then
        res := cle + cre;
    elif cle[1] = cre[1] and cle[2] = -cre[2] mod p then
        res := 0;
    else
        if cle[1] =· cre[1] mod p and cle[2] = cre[2] mod p
        then
            lambda := ((3*cle[1]^2+c)/2/cle[2]) mod p;
        else
            lambda := (cre[2]-cle[2])/(cre[1]-cle[1]) mod p;
        fi:
```

```
      x3 := (lambda^2-cle[1]-cre[1]) mod p;
      y3 := (lambda*(cle[1]-x3)-cle[2]) mod p;
      res := [x3, y3];
   fi:
   res;
end:



elgamal := proc(alpha, e, c, p)
   local calpha, n, y;
   read(addec);
   calpha := alpha;
   n := e;
   y := 0;
   while n > 0 do
      if irem(n, 2, 'n') = 1 then
         y := addec(calpha, y, c, p):
      fi:
      calpha := addec(calpha, calpha, c, p):
   od:
   y;
end:
```

C.4   Chapter 9 Procedures

```
switch := proc(n, x, maxsub)
   local vs, i, j, k, pg, bk, nsw, pe, bki, pn, allpoly,
         mon, nlist, dg, vres, colist, pnum, part, pgel,
         jnum, vt, pct, multiplicity, m;
   vs := linalg[vector](n, 0);
   vt := linalg[vector](n, 0);
   nsw := 2^n;
   read(ppoly);
   multiplicity := proc(y, j)
      j[y] := j[y] + 1;
   end:
   allpoly := 0;
   nlist := {};
   pg := []:
```

```
colist := []:
for pnum from 1 to combinat[numbpart](n) do
    for i from 1 to n do
        j[i] := 0:
    od:
    if pnum = 1 then
        part := combinat[firstpart](n);
    else
        part := combinat[nextpart](part);
    fi:
    map(multiplicity, part, 'j'):
    pgel := [];
    pct := 0;
    for i from 2 to n do
        for jnum from 1 to j[i] do
            pgel := [op(pgel),
            [seq(pct + (jnum-1)*i + k, k = 1..i)]];
        od:
        pct := pct + i*j[i];
    od:
    pg := [op(pg), pgel];
    colist := [op(colist),
    product(1/('k'^j['k']*j['k']!), 'k' = 1..n)];
od:
m := 1;
for i from 1 to nops(pg) do
    pe := pg[i];
    nlist := {};
    mon := 1;
    dg := 0;
    for j from 0 to nsw-1 do
        bk := convert(j, base, 2);
        bki := linalg[vectdim](bk);
        for k from 1 to n do
            vs[k] := 0;
        od;
        for k from 1 to bki do
            vs[k] := bk[k];
        od:
        for k from 1 to linalg[vectdim](vs) do
            vt[linalg[vectdim](vs)-k+1] := vs[k];
        od:
        vres := ppoly(pe, vs, n, x, nlist, m);
```

```
            pn := vres[1];
            nlist := nlist union vres[2];
            dg := dg + vres[3];
            m := vres[4];
            mon := simplify(mon*pn);
        od:
        mon := colist[i] * mon * x[1]^(2^n-dg);
        allpoly := simplify(allpoly + mon);
    od:
    maxsub := m;
    RETURN(allpoly);
end:




ppoly := proc(pe, vb, n, x, nlist, max)
    local i, j, dcycle, clen, ob10, nb10, res, cyct, vs,
          vc, plist, k, dg, nsum, tmp, m, ct, tmax;
    vs := [];
    vc := [];
    plist := {};
    tmax := max;
    for i from 1 to n do
        vs := [vb[i], op(vs)];
        vc := [vb[i], op(vc)];
    od:
    res := 1;
    dg := 0;
    cyct := 0;
    if linalg[vectdim](pe) = 0 then
        res := res * x[1];
        dg := dg + 1;
    fi:
    if linalg[vectdim](pe) <> 0 then
        ob10 := convert([seq(vs[linalg[vectdim](vs) - ct + 1],
                ct = 1 .. linalg[vectdim](vs))], base, 2, 10);
        if linalg[vectdim](ob10) > 1 then
            m := linalg[vectdim](ob10);
            nsum := 0;
            for i from 1 to m do
                nsum := nsum + ob10[m-i+1]*10^(m-i);
            od:
            ob10 := subsop(1 = nsum, ob10);
```

```
      fi:

if linalg[vectdim](ob10) = 0 then
   res := res*x[1];
   dg := dg+1;
   plist := plist union {0};
else
   if (member(ob10[1], nlist) = false) and
      (linalg[vectdim](pe) <> 0)
   then
   plist := plist union {ob10[1]};
   nb10 := -1;
   cyct := 0;
   while nb10 <> ob10[1] do
       cyct := cyct + 1;
       for i from 1 to linalg[vectdim](pe) do
           dcycle := pe[i];
           clen := linalg[vectdim](dcycle);
           for j from 1 to clen-1 do
               vs :=
               subsop(dcycle[j+1]= vc[dcycle[j]], vs);
           od;
           vs :=
           subsop(dcycle[1] = vc[dcycle[clen]], vs);
           for k from 1 to n do
               vc := subsop(k = vs[k], vc);
           od:
       od:
       plist := plist union {nb10};
       if linalg[vectdim](convert
           ([seq(vs[linalg[vectdim](vs) - ct + 1],
               ct = 1 .. linalg[vectdim](vs))],
               base, 2, 10)) > 1
       then
           nsum := 0;
           tmp := convert
               ([seq(vs[linalg[vectdim](vs) - ct + 1],
                   ct = 1 .. linalg[vectdim](vs))],
                   base, 2, 10);
           m := linalg[vectdim](tmp);
           for i from 1 to m do
               nsum := nsum + tmp[m-i+1]*10^(m-i);
           od:
```

```
            nb10 := nsum;

        else
            nb10 := convert
                    ([seq(vs[linalg[vectdim](vs) - ct + 1],
                        ct = 1 .. linalg[vectdim](vs))],
                        base, 2, 10)[1];
            fi:
        od;
        dg := dg + cyct;
        res := res*x[cyct];
        if cyct > tmax then
            tmax := cyct;
        fi:
        fi;
    fi;
    fi;
    RETURN(res, plist, dg, tmax);
end:
```

Bibliography

[1] Bressoud, D., *Factorization and Primality Testing*, Springer-Verlag, New York, 1989.

[2] Certicom Corp. *Current public-key cryptographic systems*, 1997, Certicom white paper.

[3] Certicom Corp., Elliptic curve groups over F2m, 1997, ECC tutorial.

[4] Guichard, D., Counting non-isomorphic graphs with Maple, *MapleTech*, (Number 8):52-56, 1992.

[5] Heal, K. M., Hansen, M. L., and Rickard, K. M., *Maple V Learning Guide*, Springer-Verlag, New York, 1998.

[6] Herstein, I. N., *Abstract Algebra*, Macmillan, New York, 1986.

[7] Hill, L. S., Cryptography in an algebraic alphabet, *American Math. Monthly*, 36:306-312, 1929.

[8] Hill, L. S., Concerning certain linear transformation apparatus of cryptography, *American Math. Monthly*, 38:135-154, 1931.

[9] Hungerford, T., *Algebra*, Springer-Verlag, New York, 1989.

[10] Klima, R. E., Applying the Diffie-Hellman key exchange to RSA, *UMAP*, Volume 20(No. 1):21-27, 1999.

[11] Koblitz, N., *A Course in Number Theory and Cryptography*, Springer-Verlag, G.T.M., 1987.

[12] Levine, J., Variable matrix substitution in algebraic cryptography, *American Math. Monthly*, 65:170-179, 1958.

[13] Lidl, R. and Neiderreiter, H., *Introduction to Finite Fields and their Applications*, Cambridge U. Press, New York, 1986.

[14] Mackiw, G., *Applications of Abstract Algebra*, John Wiley and Sons, New York, 1985.

[15] Marlin, J. H. and Kim, H., Calculus I with Maple V, Maple Calculus supplement, 1994.

[16] Monagan, M. B., Geddes, K. O., et al, *Maple V Programming Guide*, Springer-Verlag, New York, 1998.

[17] Murray, B. C., *Journey into Space*, Norton, New York, 1989.

[18] Pretzel, O., *Error Correcting Codes and Finite Fields*, Oxford U. Press, New York, 1992.

[19] Roberts, F., *Applied Combinatorics*, Prentice-Hall, Englewood Cliffs, NJ, 1984.

[20] Rosen, K., *Elementary Number Theory and its Applications*, Addison-Wesley, Reading, MA, 1988.

[21] Ryser, H. R., Combinatorial mathematics, *The Mathematical Association of America*, (No. 14), The Carus Mathematical Monographs.

[22] Sigmon, N. P., Applications of Maple to algebraic cryptography, *Mathematics and Computer Education*, Volume 31(No. 3):220-229, 1997.

[23] Sigmon, N. P. and Stitzinger, E. L., Applications of Maple to Reed-Solomon codes, *MapleTech*, Volume 3(No. 3):53-59, 1996.

[24] Stinson, D., *Cryptography, Theory and Practice*, CRC Press, Boca Raton, 1995.

[25] Tucker, A., Polya's enumeration formula by example, *Math. Magazine*, pages 248-256, 1974.

[26] Walker, G. A., *Introduction to Abstract Algebra*, Random House, New York, 1987.

[27] Waterloo Maple Inc., *Maple V Release 3 Student Version*, 1994.

[28] Waterloo Maple Inc., *Maple V Release 5*, 1997.

[29] Wicker, S. B. and Bhargava, V. K., editors, *Reed-Solomon Codes and Their Applications*, IEEE Inc., 1994.

Hints and Solutions to Selected Written Exercises

Chapter 1

4. $(12)(34)$, $(13)(24)$, $(14)(23)$, (123), (132), (124), (142), (134), (143), (234), (243), identity.

6. (12345), (13524), (14253), (15432), $(25)(34)$, $(13)(45)$, $(15)(24)$, $(12)(35)$, $(14)(23)$, identity.

7. A_4 and $(12)A_4$.

9. (a) 5

(b) 5

(c) 2

(d) 6

(e) 6

11. Let a be a cyclic generator for G, and suppose j is the smallest positive integer for which $a^j \in H$. Use the fact that Z is a Euclidean domain to show that a^j is a cyclic generator for H.

13. Example 1.7: A_n.
Example 1.8: The set of matrices A with $\det(A) = 1$.

15. Let $a \in S_n$ and $b \in A_n$, and argue that $a^{-1}ba \in A_n$.

20. Yes. Use the fact that $F[x]$ is a Euclidean domain.

23. The primes.

25. (a)

Power	Field Element
x^1	x
x^2	$2x + 1$
x^3	$2x + 2$
x^4	2
x^5	$2x$
x^6	$x + 2$
x^7	$x + 1$
x^8	1

(c) $f(x) = (x + 5)(x + 7)$ in $Z_{11}[x]$.

29. $f(x)$ is irreducible but not primitive since the order of x is 5; $g(x)$ is not irreducible since 1 is a root of $g(x)$, and $h(x)$ is primitive.

35. $(a, b) = x^2 + 1$, $u = x + 1$, and $v = x^2 + x + 1$.

Chapter 2

2. Use Propositions 2.8 and 2.9 with $p = 13$ and $n = 1$. With the cyclic generator 2 for Z_{13}^*, Proposition 2.8 yields the initial blocks $D_0 = \{1, 3, 9\}$ and $D_1 = \{2, 6, 5\}$. The parameters for the resulting block design are $(13, 26, 6, 3, 1)$.

4. Use Proposition 2.9 with $p = n = 5$. In this block design, there are 150 drivers, each car is driven 24 times, and each pair of cars is driven by the same driver 3 times. Let x be a cyclic generator for the set of nonzero elements in a finite field of order 25, and construct 6 initial blocks with 4 elements in each one. For example, the first two initial blocks are $D_0 = \{x^0, x^6, x^{12}, x^{18}\}$ and $D_1 = \{x^1, x^7, x^{13}, x^{19}\}$.

Chapter 3

2. The Hadamard code with $m = 4$ satisfies the stated requirements.

5. The following generator matrix G and parity check matrix H are one of many correct answers.

$$G = \begin{bmatrix} 1 & 1 & 1 & 0 & 0 & 0 & 1 & 1 & 1 \\ 0 & 0 & 0 & 1 & 1 & 1 & 1 & 1 & 1 \end{bmatrix}$$

$$H = \begin{bmatrix} 0 & 0 & 0 & 1 & 1 & 0 & 0 & 0 & 0 \\ 1 & 0 & 1 & 0 & 0 & 0 & 0 & 0 & 0 \\ 1 & 1 & 0 & 0 & 0 & 0 & 0 & 0 & 0 \\ 0 & 0 & 0 & 1 & 0 & 1 & 0 & 0 & 0 \\ 1 & 0 & 0 & 1 & 0 & 0 & 0 & 0 & 1 \\ 1 & 0 & 0 & 1 & 0 & 0 & 0 & 1 & 0 \\ 1 & 0 & 0 & 1 & 0 & 0 & 1 & 0 & 0 \end{bmatrix}$$

7. r_1 can be corrected to (11100), r_2 can be corrected to (11011), and r_3 cannot be corrected.

Chapter 4

1. $g(x) = p(x)$, which yields a $[7, 4]$ BCH code.

2. (a) r can be corrected to (1001110).

5. Refer to Example 4.3. Note that if we consider only the first four powers of a, then $g(x) = m_1(x)m_3(x)$, which has degree 8. The resulting code has $2^7 = 128$ codewords and is 2-error correcting.

8. (a) r can be corrected to (000111011001010).

   (b) r can be corrected to (111100010011010).

Chapter 5

1. (a) $g(x) = (x-a)(x-a^2)(x-a^3)(x-a^4) = x^4 + a^3x^3 + x^2 + ax + a^3$

   (b) The following polynomial is one of the codewords in C.

   $$(a^4x + a^5)g(x) = a^4x^5 + a^4x^4 + a^2x^3 + a^2x + a$$

   (c) The codeword above converts to the following binary vector.

   $$(010001000001011011000)$$

2. (a) $r(x)$ can be corrected to $a^5x^6 + a^6x^4 + ax^2 + a^6x + a^5$.

   (c) $r(x)$ can be corrected to $x^6 + ax^5 + a^6x^3 + a^3x^2 + a^5x + a^4$.

4. (a) $r(x)$ can be corrected to $a^7x^{12} + a^2x^{11} + a^8x^{10} + a^6x^9 + x^8 + ax^7 + a^{10}x^6 + a^6x^5 + a^4x^4 + a^7x^3 + ax^2 + a^6x + a$.

Chapter 6

3. The following is the key matrix A for the system.

$$A = \begin{bmatrix} 5 & 21 \\ 9 & 12 \end{bmatrix}$$

5. (a) "HFXLKQOOFS".

    (b) "NONETOSEND".

8. One possible way to find K is to use the 2×1 matrix

$$A = \begin{bmatrix} 2 \\ 1 \end{bmatrix}$$

and the 1×2 matrix

$$B = \begin{bmatrix} 1 & 3 \end{bmatrix}$$

to form the following 3×3 involutory matrix K.

$$K = \begin{bmatrix} 4 & 1 & 3 \\ 20 & 25 & 20 \\ 23 & 25 & 24 \end{bmatrix}$$

Chapter 7

1. (a) The ciphertext is 0 222 222 0 128 175 250 35 118 28 222
        201 99 0 216 175.

2. The corresponding decryption exponent is $b = 41$.

5. 22 total multiplications.

8. $p = 509$ and $q = 631$.

Chapter 8

1. $y = 16$ and $z = 5$.

4. $w = 2x + 1$.

7. (a) $E = \{\,(1,\pm5),\,(2,\pm1),\,(5,\pm5),\,(7,\pm4),\,(0,0),\,(3,0),\,(8,0),\,\overline{O}\,\}$.

   (b) E is not cyclic. Theorem 8.3 states then that E is isomorphic to $Z_6 \times Z_2$.

8. $y = (0,1)$ and $z = (8,2)$.

10. $w = (7,4)$.

Chapter 9

1. (c) $f(x_1, x_2, x_3) = \frac{1}{6}(x_1^3 + 3x_1x_2 + 2x_3)$, 4 distinct necklaces.

   (d) $R^3 + R^2W + RW^2 + R^3$.

3. (a) 8 distinct necklaces.

   (b) 2 distinct necklaces.

6. See Example 9.5 and the results obtained in Section 9.5.

7. 3984 distinct equivalence classes.

Index